# Columbia University

# Contributions to Education

## Teachers College Series

## No. 39

AMS PRESS
NEW YORK

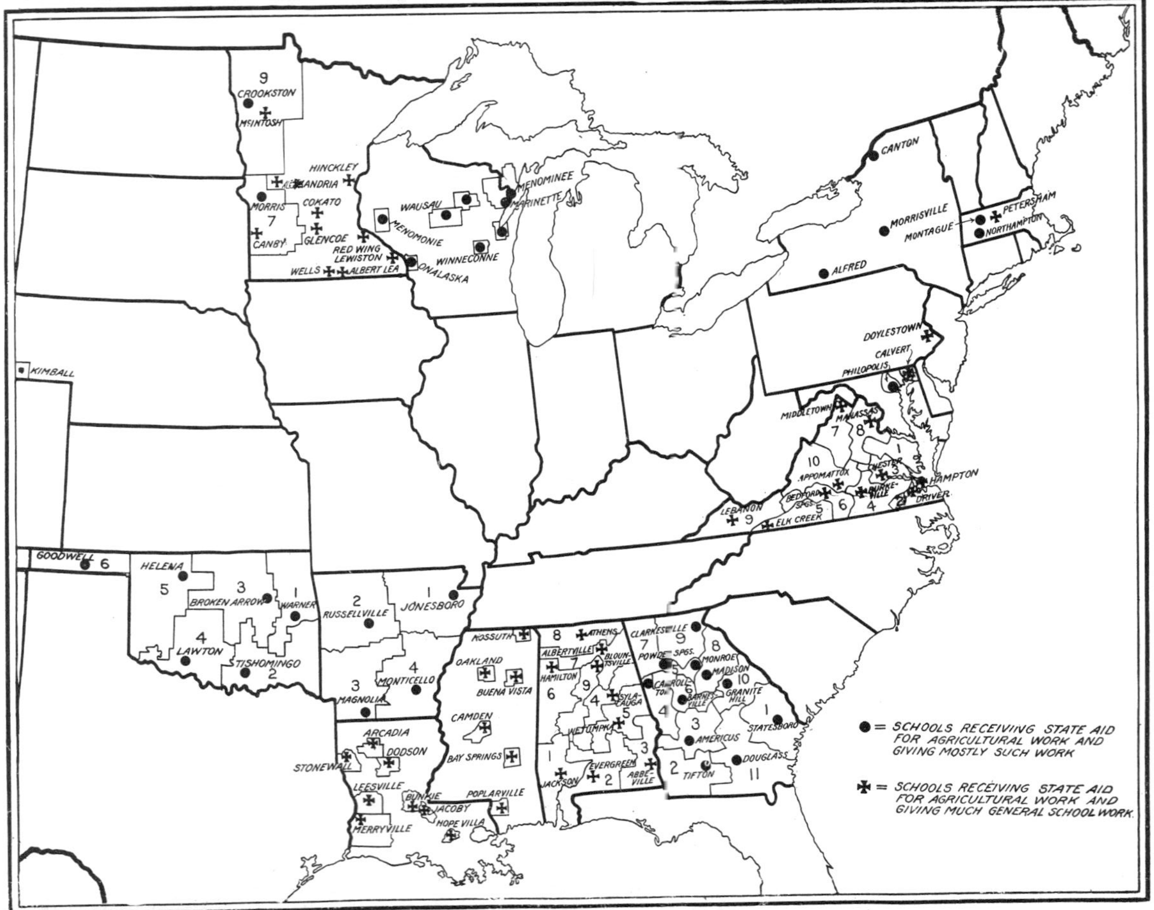
CROOKSTON
McINTOSH
HINCKLEY
MORRIS
COKATO
GLENCOE
CANBY
RED WING
LEWISTON
WELLS
ALBERT LEA
ONALASKA
MENOMONIE
WAUSAU
WINNECONNE
MENOMINEE
MARINETTE
CANTON
MORRISVILLE
MONTAGUE
ALFRED
PETERSHAM
NORTHAMPTON
DOYLESTOWN
CALVERT
PHILOPOLIS
MIDDLETOWN
MANASSAS
APPOMATTOX
CHESTER
BEDFORD SPGS.
BURKEVILLE
HAMPTON
DRIVER
LEBANON
ELK CREEK
KIMBALL
GOODWELL
HELENA
BROKEN ARROW
WARNER
LAWTON
TISHOMINGO
RUSSELLVILLE
JONESBORO
MONTICELLO
MAGNOLIA
KOSSUTH
OAKLAND
BUENA VISTA
CAMDEN
BAY SPRINGS
POPLARVILLE
ARCADIA
DODSON
STONEWALL
LEESVILLE
BUNKIE
JACOBY
HOPE VILLA
MERRYVILLE
ATHENS
ALBERTVILLE
BLOUNTSVILLE
HAMILTON
SYLACAUGA
WETUMPKA
EVERGREEN
ABBEVILLE
JACKSON
CLARKESVILLE
POWDER SPGS.
MONROE
MADISON
CARROLLTON
BARNESVILLE
GRANITE HILL
AMERICUS
STATESBORO
TIFTON
DOUGLASS
● = SCHOOLS RECEIVING STATE AID FOR AGRICULTURAL WORK AND GIVING MOSTLY SUCH WORK
✠ = SCHOOLS RECEIVING STATE AID FOR AGRICULTURAL WORK AND GIVING MUCH GENERAL SCHOOL WORK.

# AGRICULTURAL INSTRUCTION IN THE PUBLIC HIGH SCHOOLS OF THE UNITED STATES

BY

CLARENCE HALL ROBISON, Ph.D.

SOMETIME FELLOW IN EDUCATION, TEACHERS COLLEGE
COLUMBIA UNIVERSITY

TEACHERS COLLEGE, COLUMBIA UNIVERSITY
CONTRIBUTIONS TO EDUCATION, NO. 39

PUBLISHED BY
Teachers College, Columbia University
NEW YORK CITY
1911

**Library of Congress Cataloging in Publication Data**

Robison, Clarence Hall, 1873-1952.
Agricultural instruction in the public high schools of the United States.

Reprint of the 1911 ed., issued in series: Teachers College, Columbia. Contributions to education, no. 39.
Originally presented as the author's thesis, Columbia.
Bibliography: p.
1. Agriculture--Study and teaching (Secondary)--U.S. I. Title. II. Series: Columbia University. Teachers College. Contributions to education, no. 39.
S533.R6 1972 630'.712'73 75-177198
ISBN 0-404-55039-8

Reprinted by Special Arrangement with Teachers College Press, New York, New York

From the edition of 1911, New York
First AMS edition published in 1972
Manufactured in the United States

AMS PRESS, INC.
NEW YORK, N.Y. 10003

## ACKNOWLEDGMENTS

The present study would have been impossible but for the hearty coöperation of a large number of persons, as state superintendents, high-school inspectors, officers of administration or instruction in the state universities and agricultural colleges, and officers and teachers in the different schools studied. For the mass of detailed information used in Chapters II, III, IV, part of V, and VII, I am particularly indebted to the superintendents, principals, and teachers in the schools listed. (See footnote, 1.) The compilation of the list itself was rather difficult in the early stages, since agricultural instruction was so new in 1907 and because few states had at that time undertaken to collect information regarding the work.

Especially am I indebted to Mr. D. J. Crosby, Specialist in Agricultural Education of the United States Department of Agriculture, for the free use of the materials and facilities at the command of the office of Experiment Stations.

Much appreciated assistance in reading the manuscript and in checking up the tabulations has been rendered both by Mr. Crosby and by Mr. F. W. Howe, then Assistant in Agricultural Education in the Office of Experiment Stations, and now Supervisor of Agricultural Education for the New York State Education Department.

The present investigation has been carried on under the auspices of the department of secondary education, Teachers College, Columbia University. I owe a great debt to Professor Julius Sachs, head of the department, and to Dr. David S. Snedden, now commissioner of education for Massachusetts, for their helpful direction and kind suggestions at all stages of the work.

C. H. R.

# CONTENTS

## CHAPTER V

### Preparation and Salaries of Teachers of Agriculture in the High Schools



## CHAPTER VI

### Special Secondary Schools of Agriculture



## CHAPTER VII

### Problems of Agricultural Instruction in the Secondary School



## APPENDIX A



## APPENDIX B

## TABLES

# AGRICULTURAL INSTRUCTION IN THE PUBLIC HIGH SCHOOLS

## INTRODUCTION

The secondary school,[1] especially the public high school, owes a duty to the large majority of its students who do not go to college. Its problems, therefore, must be studied as problems concerned with the future of its students in the community. Just as the city high school is seeking to an increasing extent to adjust itself to present industrial conditions, so is there a rapidly growing movement on the part of the rural and village high schools to bring their work into intimate connection and close sympathy with the life and interests of their environment.

An examination of some of the features of this movement at its present stage is the main purpose of this study. It aims to furnish data much needed for that discussion which has had to depend too largely upon individual experience for its basis. Too often has personal enthusiasm been compelled to make up for the deficiency of facts. Besides giving as accurate a picture as possible of the present agricultural work in the high schools, an attempt has been made to determine the relation of this work to the school organization, to important local industries, and to the distribution of the rural population. An inquiry has been made into the present preparation of the teachers in charge of the work and the opportunities offered them to become better equipped to carry it on. An investigation has also been carried on regarding the available supply of men trained along agricultural lines, and the conditions, professional and financial, tending to limit the supply that might otherwise

[1] The schools dealt with in this investigation are all listed in the pamphlet. Institutions giving instruction in agriculture, Office of Experment Stations, October 17, 1908, p. 10. This study deals with practically all of the high schools listed on pages 3–8 of the pamphlet.

be available. From time to time some of the more important inferences are pointed out, while in the last chapter a consideration is given to some of the more pressing problems.

Among these problems may be mentioned the following: The need of more clearly defined ideas of educational principles involved and the lack of a definite method of teaching the subject; the relation of agriculture as a branch of the curriculum to the sciences already included in it; the question of duplication between the work of the high school and of the elementary grades; the attitude of the state universities outside the colleges of agriculture; the bearing on agricultural teaching of the idea of "differentiating school work at the age of twelve"; and the social implications involved in the special or technical agricultural high school for large political units. Special secondary schools of agriculture involve an expenditure of large sums of money and the selection of a number of specially trained teachers; consequently they are planned, and will be, for some time to come, after careful consultation with the scientific and agricultural experts of the United States Department of Agriculture, the state agricultural colleges, and the state department of public instruction. But this expert advice has not been taken advantage of to the same extent at least by the general, or non-specialized public high schools of the small cities and rural communities. Many of these schools have come to light in the present investigation which were unknown to some of the public agencies just mentioned and sometimes to all of them. The high schools of this latter type introduce agricultural instruction because they have faith in it, but they are hampered by lack of facilities, time, and that experience or training on the part of the teacher necessary to enable him to use the means at hand.

It is in the hope that this study may serve in some measure as a clearing house of ideas on these points that it is largely devoted to an inquiry into the present status of agricultural instruction in this type of schools. Some studies of other types of schools are introduced for the help they may contribute to an intelligent discussion of the non-urban high-school problem, and are not intended to be exhaustive presentations of the work of those institutions.

The source material has been, first, direct data on the high schools themselves: (1) Questionaire replies from superintendents, principals, and special teachers of agriculture; (2) catalogues, class material, and other records furnished by these schools; (3) personal visits to selected schools scattered from the Atlantic Coast to the Missouri River; and (4) some little material furnished by the state reports. Second, contributory data: (1) Documentary data and personal visits to summer schools attempting to help the high-school teachers; (2) returns from officers of administration and graduates of agricultural colleges, bearing on the salary question; (3) data from special agricultural schools, bearing on certain phases of the general question; and (4) special reports of the United States Department of Agriculture, of the Bureau of Education, of state boards of agriculture, of the National Education Association, various addresses, and miscellaneous sources.

So far as possible, the replies have been reduced to similar terms so as to make comparable as large a number of schools as possible. But owing to omissions of certain items in the replies, the number of such comparable cases is considerably smaller than the number even of the fuller replies, ranging from fifty per cent to ninety per cent, according to the items compared. It has been possible to supply estimates of population, enrollment, etc., in no small degree from the state reports of Nebraska and Ohio, which states include over one-half of the schools reporting.

# CHAPTER I

## AGRICULTURAL EDUCATION

The Committee on Industrial Education in Schools for Rural Communities[1] expresses the opinion that "industrial education has for its purpose the acquiring of a body of usable knowledge of greater or lesser extent relating to industrial conditions, processes, and organization, and to the administration of affairs incident to the environment of the individual being educated, involving the gaining of some skill in the use of such knowledge, and the securing of mental, aesthetic, and ethical training through the acquisition and use of the knowledge indicated." The members of the committee were L. D. Harvey, chairman, L. H. Bailey, Alfred Bayliss, W. T. Carrington, and W. M. Hays. To make the above statement apply exclusively to agricultural education it is only necessary to qualify appropriately the word environment. In a narrow and more formal way, agricultural *education* might be regarded as the mastery of the principles underlying farm practice. Agricultural *training*, on the other hand, is the gaining of considerable skill in carrying on farm operations. It may be and usually is obtained by imitation or by following rules derived the user knows not how. A proper mastery of principles, however, involves more or less participation in the operations, used solely as laboratory exercises, whether carried on indoors or out.

### A Brief Historical Sketch

One of the earliest proposals in this country to regard agriculture as a fit subject for higher education is found in a prospectus issued by William Smith, in 1751, designed as a model

[1]Report of the committee in the National Education Association, Journal of proceedings and addresses, July, 1905, p. 10.

for colleges.[2] This plan, providing for the chemistry of agriculture, was carried out more or less fully at Philadelphia Academy (University of Pennsylvania). We find Husbandry and Commerce mentioned in the original prospectus of King's College (Columbia University), dated May 31, 1754, and Agriculture and Merchandize in the laws and orders adopted by the governors, June 3, 1755.[3] The chair of botany and agriculture in 1792 was held by Samuel Latham Mitchell, M. D. In 1794, in describing a summer course in botany, he says "An attempt is made by the professor, who is a practical farmer, to elucidate and explain the economy of plants, their affinity to animals, and the organization, excitability, stimuli, life diseases, and death of both classes of beings. The physiology of plants, . . . is therefore particularly enlarged upon, as connected with gardening and farming."[4]

One of the best instances of the actual uses of agriculture and other industrial work in an educational way for pupils of elementary and secondary school age is furnished by the schools established at New Harmony, Ind., in 1825, by William Maclure,[5] in connection with his socialistic experiment known as the New Harmony Movement. Maclure placed the schools in charge of Joseph Neef, whom he had brought to Philadelphia in 1806 to introduce Pestalozzi's method of teaching. He provided ample dormitories, books, museums, shops, experimental plats, and other facilities. The experiment was short-lived, suffering from the spirit of religious intolerance on all sides, while the location so far from the older centers of intellectual life was largely responsible for the slight impression the schools made on educational practice.

A pioneer movement in agricultural education and one that lasted much longer than many others, though not much noticed

[2] William Smith, Discourses on Public Affairs, second edition, London, 1762.

[3] Van Amringe *in* Universities and their Sons, edited by Joshua L. Chamberlain, pp. 583 and 598.

[4] Mitchell, The Present State of Learning in the College of New York. New York, 1794.

[5] Will S. Monroe, Pestalozzian Movement in the United States.

in the literature of agricultural education, was the Oneida Manual Labor Institute, conducted by George Washington Gale from 1827 to 1834, and including instruction in carpentry and agriculture. This extended effort followed a few years' experience with a number of boys who were taken on to a farm near Whitesboro, Oneida County, N. Y., to which he had retired from the ministry on account of ill health. He later established Knox College, at Galesburg, Ill.

A Manual Labor Academy was conducted from 1830 to 1832 at Germantown, Pa., by George Junkin, who was later the first president of Lafayette College.

Sporadic attempts, more or less futile, were made to introduce regular instruction in agriculture as a part of the school curriculum early in the last century, as at Dummer Academy, Newberry, Mass., 1824; Derby, Conn., 1824; Teachers' Seminary, Andover, Mass., 1838; The Peoples' College, Montour Falls, New York, 1853; Westfield (Mass.) Academy, 1856; and Powers Institute, Bernardston, Mass., 1857. The opening of Bussey Institute, founded by a bequest made in 1842, was delayed by Harvard College until 1870. That commendable philanthropy, the Farm School, Thompson's Island, Boston, now ninety-five years old, began instructions in agriculture in 1833, and has since continued it. Work of recognized scientific merit was inaugurated at the Sheffield Scientific School, Yale, in 1848, by the establishment of a chair of agricultural chemistry and vegetable and animal physiology.

Most of the present large list of agricultural and mechanical colleges were founded as a result of the famous Morrill Act of 1862. Of the few already in existence before this date, the Michigan Agricultural College, opened to students in 1857, is the oldest. The Morrill Act gave to each state 30,000 acres of land for each member of Congress for the establishment and maintenance of such schools. From this have resulted endowment funds amounting to $12,000,000, with $8,000,000 worth of land not yet sold. Later acts of Congress, the Hatch Act effective in 1887; Morrill Act, 1890; Adams Act, 1906; and Nelson Act, 1907, have appropriated sums of regularly increasing amounts which in 1912 aggregate $80,000 annually for the agricultural colleges and experiment stations of each state.

By June 30, 1909, the end of the fiscal year, there had been paid to the land-grant colleges by the Federal Government an approximate total of $23,000,000.

Although the colleges for several years derived their principal support from the income of the grants of land, the Morrill and later acts have proved such a spur to the states that a large number appropriate sums each year amounting to many times the aid received from the federal treasury.

This is strikingly shown by the fact that the total federal aid given the land-grant colleges for all purposes for the fiscal year ending June 30, 1909, amounted to $2,641,006.63 while the total income from state appropriations amounted to $10,172,-559.48, making a grand total of $12,813,566.11.[6]

An agricultural college exists in every state and territory except Alaska either separately,[7] in connection with the state university,[8] or as a part of a semi-public institution.[9]

The value of all property of the land-grant colleges in June 30, 1909, was $111,882,686.96, including permanent funds amounting to $34,285,131.71, and their total income was $18,-082,853.55. Out of 28,686 white collegiate students, 7,038 were in agricultural courses of all kinds or home economics, while 10,409 were in corresponding "short courses."[10] The enrollment in the engineering courses increased 14 per cent from

[6] Office of Experiment Stations Report, 1909.

[7] Alabama, Colorado, Connecticut, Hawaii, Indiana, Iowa, Kansas, Massachusetts, Michigan, Mississippi, Montana, New Hampshire, Nevada, New Mexico, North Carolina, North Dakota, Oklahoma, Oregon, Pennsylvania, Rhode Island, South Carolina, South Dakota, Texas, Utah, Virginia, Washington—26.

[8] Arizona, Arkansas, California, Florida, Georgia, Idaho, Illinois, Kentucky, Louisiana, Maine, Minnesota, Missouri, Nebraska, Ohio, Tennessee, West Virginia, Wisconsin, Wyoming,—18.

[9] Delaware, Maryland, New Jersey, New York, Vermont,—5.

[10] Office of Experiment Stations Report, 1908, p. 192. Does not include the few colored students.

1906 to 1907, while that of the agricultural courses increased 25 per cent.[11] Using comparable data in the report of the Commissioner of Education for 1908, the increase in the enrollment in the engineering and purely agricultural courses is 10 per cent and 18 per cent respectively.[12]

In the following year we find a decrease in the gain in both courses, with the disparity growing greater. The increase in the enrollment in the agricultural courses was 14 per cent, while the enrollment in engineering courses was practically at a standstill.[13]

The primary purpose and function of the experiment station is research, invesigating both general and local problems. These stations are attached to the state agricultural colleges except in Ohio and Georgia, where they are separate and in different localities. Separate stations also exist in many states for the study of special problems, and there are stations in Alaska, Hawaii, Porto Rico and Guam. When established as departments of the colleges, part of the time of the staff is available for instruction.

The disbursements from the United States Treasury to the states and territories for agricultural experiment stations under the so-called Hatch and Adams acts of 1887 and 1906 amounted for the year ending June 30, 1909, to $1,248,000.00.[14]

## Present Agencies of Agricultural Education

The earliest organized form of agricultural education is to be found in the farmers' clubs and societies started in the older states at a very early day, and later in the arrangements now generally known as farmers' institutes, by which speakers were provided through a central authority, either for these local societies, or for general public meetings. We find the Pennsylvania Society for Promoting Agriculture as early as 1785.

[11] Report of the United States Commissioner of Education, 1907, p. 871.
[12] Ibid., 1908, vol. 2, p. 738.
[13] Ibid., 1909, vol. 2, pp. 1012–13.
[14] Report of Office of Experiment Stations, 1909, p. 226.

The Massachusetts Society for Promoting Agriculture tried, as early as 1796 and 1800, to act as a medium for the exchange of ideas and discoveries in agriculture and for their spread throughout the state. Stock exhibits in the state gave rise to various societies in the first decade of the nineteenth century. These societies were responsible for the efforts to introduce agriculture into the Massachusetts schools already mentioned. In New York almost from the beginning of the century to the establishment of the College of Agriculture at Cornell in 1868, speakers were furnished to local clubs by the State Agricultural Society. The first agricultural society in New Hampshire, so far as known, worked under a charter granted in 1814. The county societies received state aid from 1817 to 1820. The Maine legislature in 1832 voted premiums to the various agricultural societies; and the required reports of the exhibitors were printed in the *Maine Farmer,* and were discussed by the farmers' clubs throughout the state.

State boards of agriculture, organized under one name or another toward the middle of the century and after, have served as central bodies to direct farmers' institutes, manage state fairs, carry on investigations and tests, and protect farm interests from various forms of pests and injury. The 44 states and territories reporting for the year ending June 30, 1909, show an aggregate of 15,535 half-day sessions in 5,014 institutes, addressed by 1,130 state speakers and possibly three times as many more local speakers. The cost of these institutes amounted to $328,660.86.[15] Almost 40 per cent of the state lecturers hold college degrees, and another 16 per cent have taken partial college courses.[16]

Since 1903, federal aid, other than financial, has been extended to the state institutes through a "farmers' institute specialist," attached to the Office of Experiment Stations.

Akin to these are the various activities of the agricultural

[15] Yearbook, Department of Agriculture, 1909, p. 137.

[16] Based on data in History of Farmers' Institutes in the United States, by John Hamilton, Bull. 174, Office of Experiment Stations, p. 7.

colleges. They conduct "short courses" of from one week to several months, during the winter, which are open to all, offer correspondence courses, and run traveling schools, such as the so-called "corn specials," "alfalfa trains," etc. This unique form of school is a train of several cars fitted up as small museums, lecture rooms, and quarters for the instructors. Stops are made at small stations and cross-roads on the railroad according to an advertised schedule for a short time varying from several minutes to a few hours. Another form of college extension work is the boys' club contests for the exhibition of corn, cotton, or other produce of local importance, held in the various counties of the state, while the corresponding girls' home economics clubs make appropriate displays. The University of Illinois inaugurated the plan of having the prizes take the form of trips to the university during the winter short course.

While the government funds are distributed to the agricultural colleges by the Department of the Interior, the Department of Agriculture acts as the great clearing house for the activities of the colleges and experiment stations, prosecutes extensive scientific investigations at home and abroad, and seeks to control or lessen disasters of interstate importance to agricultural interests, such as the cotton-boll weevil and the foot-and-mouth disease. These various reports and findings are issued as regular or special publications of the department.

No less wonderful than the development of the agricultural college has been the rapid growth within the present decade of the agricultural movement in the elementary and secondary schools of our public school system. Instruction in the rudiments of agriculture is required in the elementary schools of Alabama, Arkansas, Florida, Georgia, Louisiana, Maine, Mississippi, Missouri, North Carolina, Oklahoma, South Carolina, South Dakota, Texas, West Virginia, and Wisconsin, fifteen in all.[17]

---

[17] From reports of state superintendents to the author and the United States Department of Agriculture. See Annual Report Office of Experiment Stations, 1906, p. 271 and Report of the Commissioner of Education, 1919, vol. 1, pp. 276–277.

There seems to be no data at present upon which to base even an approximate estimate of the number of children receiving instruction of this sort in elementary schools.

There are many private institutions of special character offering instruction along agricultural lines. These are, for the most part, of secondary grade, or of mixed secondary and elementary grade, although some of the denominational schools do work of college grade outside of the agriculture. Some are benevolent institutions of unique character, as the Mount Hermon School, founded by D. L. Moody, at Mount Hermon, Mass.; the Baron De Hirsch School at Woodbine, N. J.; Hampton Institute; Dr. Washington's famous experiment at Tuskegee, and others. Agricultural work of some sort is coming to hold a large place in the occupational instruction of many orphanages, corrective institutions, and other charitable enterprises.[18] It is a question in the case of some of these how far the work is consciously used as an instrument for the education of higher mental processes, and how exclusively it is confined to routine manual labor incidental to the operation of lands belonging to the school.

For the purpose of this bulletin, the secondary schools may be divided roughly into the following types: (1) The general, public high school. (2) The private academy, which still survives occasionally alongside of the public high school and bids for support from the same clientage, but which more often performs the functions of a high school for the community. The schools of this type that teach agriculture may not number more than a dozen, and for many purposes of this study, may be considered as belonging in the same class with the public high school. In these non-specialized schools, the time given to agriculture may vary from a four-week "normal course" to a four-year course. The courses are usually a half or a full year in length, elective or required. (3) The technical high school of agriculture, which exists primarily to give instruction in the agricultural sciences and a certain familiarity

[18] J. R. Jewell, Agricultural Education, pp. 64, 65, 79, 83, and 84.

with the art of farming. The non-agricultural studies may vary from three-fourths to five-sixths of the time in a given term or year in some schools, down to a proportion as low as a tenth of the time in certain terms of other schools. So that the line of demarcation between these and the one extreme of the general public high school is sometimes one of name and organization only. This will be discussed more fully in Chapter VI. (4) A type of special school giving agricultural work of secondary grade, is the teachers' training school, whether maintained by the state, county, or town. The output of the state normal schools is largely absorbed by the elementary schools of cities and towns where agriculture is not taught, and where it has practically no influence on the nature-study given. The work of the state normal schools is largely important in this study for its bearing on the rural high school teacher problem, but is too large a subject to be considered in this study.

# CHAPTER II

## THE PUBLIC HIGH SCHOOL

At present there is a gap between our primary schools in country and city and the industrial collegiate courses which must be closed, and if necessary the nation must help the state to close it. Too often our present schools tend to put altogether too great a premium upon mere literary education, and therefore to train away from the farm and shop.—THEODORE ROOSEVELT.

### IMPORTANCE OF AGRICULTURAL INSTRUCTION IN THE HIGH SCHOOL

Several considerations point to the importance of agricultural instruction in high schools. One is the large dependence of the rural elementary schools upon the local high schools for their supply of teachers. The officers of 132 high schools reporting on this point indicate an approximate estimate of 1,722 pupils, out of the total enrollment of 11,977, who later teach in the country schools. Returns from 160 high schools show that 4,071 pupils, out of the total enrollment of 15,243, are from farm homes. These schools minister to a constituency of nearly half a million people. This has a bearing on the problem to whatever extent we believe these schools should relate their work closely to the interests of the community.

Attention has been called by Professor Thorndike[1] to the fact that "the most typical, in the sense of the most frequent, secondary school in the United States is a school taught by one teacher. The secondary schools in the country with only one teacher outnumber by a considerable figure all those with five or more teachers. Those with only one or two teachers outnumber by a considerable figure all the rest. Those with one, two, or three teachers are ten times as frequent as those

[1] *Educational Review*, Vol. 33, March, 1907, pp. 245–255.

with ten or more teachers and five times as frequent as those with from five up to ten teachers."[2]

The data in the report of the United States Commissioner of Education for 1904, from which he makes his computations, show that out of a total of 7,174 high schools in United States up to that time there were 2,175 high schools with but one teacher, 1,807 with two teachers, 1,221 with three teachers; or 3,982 with one or two teachers, and 5,203 with three teachers or less.

The same data also show that over 36 per cent of the high-school pupils of the United States are in these schools with three teachers or less.

## Types of Secondary Schools Teaching Agriculture

From the standpoint of the average American citizen, secondary schools giving instruction in agriculture fall into two groups: (1) Those supported by public funds, regardless of how the money is raised, and (2) schools supported by private benefactions. From the standpoint of administration, however, the line of cleavage is rather between the general, or non-specialized, public high school with agriculture included among the various other studies taught, and the special, or technical, agricultural high school. Many technical agricultural schools are private, but appealing to a general constituency, while many others, both public and private, are for mental and moral delinquents. None of these technical schools will be considered in this work except those maintained by public funds and open to the young people of the community, and then but briefly in Chapter VI.

The general high school offering instruction in agriculture is of practically every type recognized among public schools, except the large city type. The political units supporting it range from villages and parts of townships to counties. The special schools, on the other hand, are supported almost without exception by the larger political units, the county, the congressional or special district, or the state at large.

[2] *Educational Review*, vol. 33, March, 1907, table I, p. 253.

The aim of the special schools is avowedly vocational. They possess facilities for carrying on practical farming in its various branches, which are surprisingly complete in view of the short time since the inception of the idea. Where these schools have started *de novo* we find the most radical departure in the way of equipment and curriculum. Where agricultural departments have been grafted on to existing institutions, or where these have been "reorganized," there we find the least innovation. This statement, with appropriate change of phraseology, applies equally well to the public school.

We find special schools on the one hand with but few acres of land or no ground at all for practical work, and on the other hand with large tracts for the illustration of quite diversified methods of farming. Sometimes we find fair-sized observation tracts but no room for outdoor laboratory work; if the students do farm work in such schools, much of it may be as day laborers working out their board, diminishing the time for school work proper. This is not apt to be carried to such an extreme in the case of publicly supported schools as in private or denominational schools, where the underlying principles are often ignored. In the matter of curriculum, we may find at one extreme no cultural elements save a very limited amount of English, and perhaps civics, and at the other extreme we find schools established ostensibly as "agricultural high schools," or calling themselves by that name, with the usual Latin-scientific courses, with distinct agriculture running as a parallel course through three or four years. With very few exceptions, this last-named arrangement represents the most advanced position of the general high school, and at this point the characteristics of the two large groups overlap. While this was the case in some of the Alabama district schools as shown by their catalogues but two or three years back, we now find the position reversed, with agriculture required of all for five semesters and an option of substituting Latin for it during the last three semesters. This course, outlined in Chapter VI, went into effect September, 1909.

The Alabama state superintendent of education is revising the high-school course of study so as to include the subject of agriculture for county high schools established under the act of August 7, 1907. These schools receive $2,000 annually from the state. High schools have been established under these provisions in 34 of the 67 counties of the state.

County high schools in Kansas must prepare for the State Agricultural College as well as for other colleges, but reports both from the Agricultural College and the State University seem to indicate that the strong influence of the latter is thrown decidedly on the side of classical work and indirectly against agricultural work.

The parish or county "agricultural" schools now being established in Louisiana are organized under the general high-school law. They receive the same state aid, about $350 for the year 1909-10, that county high schools have heretofore been given, although a great effort is now being made to secure an appropriation of $25,000 for the county schools teaching agriculture.[3] These schools are on the same basis as other approved high schools, and do four years' work, from the eighth to the eleventh grades inclusive. For the present they will admit only boys, although domestic science courses for girls may be added later. Each school must be provided by the county with certain prescribed facilities in the way of "apparatus $300, equipment $300, land not less than 5 acres, barns, teams, etc." The course in agriculture is distinct from other high-school courses, although some schools have both literary and agricultural courses. So far as purpose and content of instruction go, those without the literary courses might strictly be classed with "agricultural" schools, while the others although supported in the same way may properly be included among the non-specialized high schools. Nine have so far been established, in Arcadia, Bunkie, Dodson, Hope Villa, Jacoby, Leesville, Merryville, Stonewall, and in one other village.

In Michigan the course of study for rural township high schools "may include instruction in manual training, domestic science, nature-study and the elements of agriculture."

The legislature of Minnesota, in 1909, provided aid for the establishment of departments of agriculture, manual training,

[3] The legislature of 1910 gave $50,000 for special work in agriculture. This will give each school $1,200 to $1,500.

and home economics, to the extent of two-thirds the amount expended by the school, but not to exceed $2,500 for each school. This aid was restricted to ten high or consolidated schools the first year, and to the same number of additional schools each succeeding two years. The school must have ample facilities, instructors qualified to teach the industrial subjects, and 5 acres of land. The entire quota was filled the first year, by ten high schools, namely: Albert Lea, Alexandria, Canby, Cokato, Glencoe, Hinckley, McIntosh, Red Wing, and Wells, and one consolidated graded school at Lewiston.

Mississippi has a law, passed in 1908, and reenacted in 1910 in form to meet constitutional objections, "An Act to provide for the establishment and equipment of county agricultural high schools and to provide for the equipment and maintenance of same." It provides that "instruction shall be given in high school branches, theoretical and practical agriculture, domestic science, and in such other branches as the board . . . may make a part of the curriculum, subject to review and correction by the state board of education." It will be seen that the degree to which these schools are purely agricultural depends somewhat upon the nature of the curriculum, one school including Latin and Greek in its projected course of study. Every school, however, must have 40 acres of land and suitable buildings, including dormitory accommodations for 40 students, before receiving the $1,500 aid from the state. The law is quoted more fully in Chapter VI.

The state of Virginia in its agricultural instruction combines the plan used in Michigan, Nebraska, and New York, of having a state-wide system of high-school normal training classes, with the Wisconsin plan of having a county normal training class housed in the building used by the agricultural school, while the geographical unit is that used by the Alabama and Georgia agricultural schools. The catalogues and letterheads of several of these schools variously bear the legend of "training school," "high school," "agricultural high school," and the ——th congressional district agricultural school," showing the lack of official sanction of any particular title. The state superintendent reports that there is "no legal designation for agricultural schools when they continue the classical or college preparatory work as before." The five schools thus reorganized by 1909 all maintain college preparatory (classical) courses, and the other four may be doing so. In some cases the schools report that the growth of the agricultural work has been impeded by the reluctance of the local patrons to allow the agriculture to be substituted for anything else, instead of its being superadded to an already full curriculum.

None of the state laws or regulations, so far as I have been able to determine, contain anything to obstruct the introduction of agricultural instruction in the high-school curriculum if it received the sanction of the proper supervisory authority. Indeed, its actual introduction into the high schools of many states is largely due to the persistent efforts of the state departments of public instruction. For example, the state departments of Indiana, Michigan, New Hampshire, and New York, have carefully planned curricula centering about the agricultural subjects for the guidance of high schools and academies. Several state universities have defined such a "unit in agriculture" as they will accept for entrance credit into courses leading to one or more degrees. The Massachusetts Industrial Commission and the Agricultural College at Amherst are performing a like service for their state, exercising close personal supervision over the work at Montague and Petersham through Professor W. R. Hart, in charge of education at the Agricultural College, and through the agents of the commission. Professor W. H. French, late of the department of public instruction of Michigan, holds a similar position at the Michigan Agricultural College, and has oversight of the effort to conduct a model agricultural high-school course at North Adams.

From the foregoing survey it will readily be seen that the term "agricultural high school" is largely a matter of definition. It would seem as logical for a high school offering Latin and a four-year commercial course to call itself a commercial high school, as for one offering Latin and a four-year agricultural course to call itself an agricultural high school. According to its course of study, the Cecil County Agricultural School, Calvert, Md., requires all of its students to take both Latin and agriculture.[4] During the year 1907-8, forty students were enrolled in the first and second years; so that it is hardly proper to offer an adverse criticism because any of the work is not elective. During the year 1908-9, one, at least, of the congressional district agricultural schools of Alabama required its students to take both Latin and agriculture.[5] Sixty pupils were enrolled in the eighth (the first year of high-school work), ninth, and tenth grades.

A school in Mississippi, organized under "An Act to provide for the establishment of county agricultural high schools," and advertised to open September, 1909, includes Latin and Greek in the second and third years of its suggested course of a study.[6]

On the other hand, the printed courses of study of the Beaverhead County High School, Dillon, Mont.; the Guthrie County High School, Panora, Iowa; and the Norton County High School, Norton, Kans., for 1907, 1908, and 1909, respectively, show that optional courses of four years each are offered in the classics, commerce, and agriculture. But these schools consider themselves only general, or non-specialized, high schools. The township high schools at Petersham, Mass., and Waterford, Pa., offer classical and agricultural "courses" of four years each, while the John Swaney School, Magnolia township, Putnam county, Ill., offers a choice between Latin and agriculture or manual training in each of the four years for the boys, and between Latin and home economics for the girls. The John Swaney and Petersham township high schools enrolled in the first and second years' work (the only ones yet in operation) only 32 and 37 pupils, respectively, in the year 1907-8, but neither made it necessary for every student to take Latin, nor did they style themselves "agricultural high schools," although the former has an 8-acre state demonstration plat, and the latter 2 acres for student experimentation. But these can hardly be classed with the 150-acre farm forming the experiment station of the school at Athens, Ala.

[4] Proceedings of the National Education Association, for 1908, p. 409.

[5] Catalogue of Eighth District Agricultural School, Athens, Ala.

[6] Catalogue of Yalobusha County Agricultural High School and Experiment Station.

It seems reasonable that the term "agricultural high school" should be restricted to an institution that requires all of its students to take agriculture, that does not require them to take the classics, or still better does not even teach the classics at all, that does require its students to devote to agriculture at least one-fourth of the entire time, that is, to give it as much attention as any other full-time subject, and that makes definite provision for practice in farm operation. It is equally important that the school should make ample provision for the other sciences. Some schools are attempting to make "agriculture" take the place of the foundation sciences. A school using this title should devote at least one-third or one-half of its time to agriculture and the related sciences. However, where institutions are designed by law as agricultural high schools, though they are really general public high schools there is no alternative but to so classify them. It is not necessary to do so in cases where state departments and local boards are appropriating the term without specific authorization, as in Louisiana and Virginia. These cases will be treated more fully later.

## Legislation

An examination of the school laws of a large number of states, most of them as late as 1907 or 1908, and a few later, fails to show many specific references to the teaching of agriculture in the general public high school. But much important legislation was enacted during the sessions of 1908, 1909, and 1910.

The state superintendent of New Jersey has ruled that agricultural instruction comes within the intent of the act of 1903, granting state aid equal to the appropriations made by the local communities for industrial instruction, from $250 to $7,500.[7] A somewhat similar provision with a lower limit of $3,000 has been in force since 1881. The remarkable thing is that no school has so far taken advantage, for agricultural purposes, of this most liberal provision of any state in the Union for industrial education. The laws of Vermont refer to the teaching of "in-

[7] Report of the committee on industrial education in schools for rural communities, National Education Association. Proceedings and addresses, 1907, p. 434.

dustrial science," and those of Maine to the teaching of the "natural sciences in their application to mechanics, manufacture, and agriculture"; but no evidence of such teaching has been found in the public schools of either state. The legislature of Maine in 1907 made an appropriation of $500 a year.

In county high schools of Nebraska "there shall be taught and practiced in the ninth and tenth grades, manual training, domestic science, and the elements of agriculture"; and the methods of teaching agriculture in the normal classes of the eleventh and twelfth grades. They must also conduct an experimental plat of at least 5 acres. The first high school established under this law opened for the year 1909-10 at Kimball, and a second has been voted.

Oklahoma requires agriculture in the "public schools."

The legislature of Texas in 1909 set aside $32,000 to duplicate sums from $500 to $2,000 appropriated by school boards to establish departments "for the teaching of agriculture, including such courses in manual training and domestic economy as are subsidiary to agriculture." Each high school applying for such aid must satisfy the state board of education that it has ample laboratory facilities and land, and that it will continue to maintain the department after the state aid, given for one year, is withdrawn.

The Virginia legislature appropriated, in 1908, $20,000 to establish "departments of agriculture, domestic economy, and manual training, in at least one high school in each congressional district."

Industrial departments have been added to the school system in Appomatox, Appomatox County; Burkeville, Nottoway County; Chester, Chesterfield County; Driver, Nansemond County; Hampton, Elizabeth City County; Lebanon, Russell County; Elk Creek, Grayson County; Middletown, Frederic County; Manasses, Prince William County; and Bedford Springs, Campbell County, supplying all ten districts. Each of these high schools has been designated to receive not over $1,500 of the $15,000 appropriated for normal training departments.

In those states where this movement is much more widespread, as in Nebraska and Missouri, the same kind of encouragement was given during the early stages. In Nebraska, where agriculture is taught in a larger number of first grade high schools than in any other state, much of the later influences can be traced to the demand for well trained rural school teachers, who come in no small degree from the city high-school training classes.

The spread of agriculture in Ohio, where it is now taught in more of the strictly rural high schools than in most of the other states considered, is due to the earnest and sympathetic campaign carried on by the State University through its agricultural extension officer, Professor A. B. Graham.

Because of the expense of hiring a competent instructor, of getting apparatus, and of maintaining an experimental plat, many leaders in the movement for agricultural instruction assume that its success depends on the establishment of high schools in the larger units, as the township, county, or even larger units, such as the congressional or judicial district. They feel that a township and village together will be better able to furnish the needed funds than either township or village alone, and that a county high school will be more sympathetic toward the movement than a high school supported by the city alone.

Legal provisions exist for the establishment and maintenance of county high schools in Alabama, California, Iowa, Kansas, Louisiana, Mississippi, Montana, Nebraska, Oregon, and Tennessee. High schools for entire townships, parts of the same, or for contiguous parts of adjoining townships, may be established in Illinois, Indiana, New Jersey, New York, North Carolina, North Dakota, Ohio, Pennsylvania, South Dakota, Wisconsin, and by the New England "towns." Other states have also various forms of "special districts."

## Statistics of Schools Teaching Agriculture as a Separate Study

Tables 1, 2, 3, and 4 give a conservative statement of the schools that are probably doing the agricultural work they claim to do, whether it is for half a year or for the entire four years. When any considerable number of schools are said to teach agriculture, either in the published reports, or in private correspondence, the number usually shrinks materially when subjected to an impartial examination of each case on its own merits. Thus one optimistic state superintendent in 1906 reported 200 high schools as teaching agriculture in his state. The report of his successor made the following year, containing the returns from a newly appointed high-school inspector, showed only 60 schools. Personal correspondence with the school principals of this state justifies the conclusion that not over 30 schools taught agriculture that was more than mere book work, without even window-sill or tomato-can experiments. Many of them reported that the subject was not taught at all. Recent estimates running into four figures do not seem at all warranted by the known facts. It must be said, however, that the number of public high schools, both general and special, teaching agriculture is increasing with marvelous rapidity. In the school year 1906-7, the number probably did not exceed 75 or 80, even including the agriculture taught in training classes attached to high schools. In 1907-8, there were probably between 240 and 250; while in 1908-9, the total had risen to the neighborhood of 500. It seems safe to say that the number is increasing at the rate of about 100 per cent each year.

Perhaps a better index of the real progress is the number of schools having in actual operation two or more years of agricultural instruction. In 1906-7 there were about 15 general and special schools giving instruction for two or more years, most of these being special schools. In 1907-8, the number had increased to 22, about one-third of which were general high schools, while several other schools were starting this more extensive work. In 1908-9, the number of high schools and academies that serve their communities as public high schools offering such work was at least 24, while 28 more receiving state aid were of mixed character, usually offering more or less classical work, but sometimes calling themselves agricultural high schools. The number of strictly agricultural secondary schools receiving state aid was close to 35. In addition to these, 9 organized the following year would be classified under the first heading, 10 under the second, and 10 under the third, including the 4 authorized in Arkansas and now organizing.

These schools are enumerated by states, and classified according to the views expressed in the preceding pages.

The second classification by states is in accordance with the legal designation or with the local or popular reputation and claims. That these claims are sometimes not warranted is apparent on examination of the curriculum, equipment, and work of a number of the schools mentioned in Table 2, and comparison with the same criteria of a number of schools in the first column of Table 1. The few private academies included in Table 1, are known to serve the purpose of public

TABLE 1

HIGH SCHOOLS AND ACADEMIES OFFERING INSTRUCTION IN AGRICULTURE FOR TWO OR MORE YEARS

| Schools depending on local support or patronage | | Schools receiving state aid and offering general high school work | | Technical schools receiving state aid for agriculture | |
|---|---|---|---|---|---|
| California | 3 | Alabama | 9 | Arkansas | *e*0 |
| Illinois | 1 | Louisiana | *c*4 | California | 2 |
| Iowa | 1 | Minnesota | *b*3 | Georgia | 12 |
| Kansas | 1 | Mississippi | 6 | Louisiana | *c*5 |
| Maryland | 1 | Oklahoma | 1 | Massachusetts | 2 |
| Massachusetts | *a*1 | Virginia | *d*5 | Michigan | 1 |
| Michigan | 7 | | | Minnesota | 1 |
| Montana | 1 | Total | 28 | New York | 3 |
| New Hampshire | 2 | | | Oklahoma | *f*2 |
| New York | 1 | Starting in 1910 | 10 | Pennsylvania | 1 |
| Pennsylvania | 1 | | | Texas | 1 |
| Tennessee | 1 | | | Wisconsin | *g*5 |
| Utah | 1 | | | | |
| Vermont | 1 | | | Total | *h*35 |
| West Virginia | 1 | | | | |
| | | | | Starting in 1910 | 10 |
| Total | 24 | | | | |
| Starting in 1910 | 9 | | | | |

*a* This school has applied for state aid.
*b* Six more will institute agricultural departments in 1910.
*c* Four of these Louisiana schools do not offer classical work, and five offer agricultural but not literary courses, but all operate under the general law.
*d* Ten in all are now organized. Many of these offer classical courses. Information at hand does not show how many offer more than one year of agricultural work.
*e* Two schools have already been located and two more will be in 1910.
*f* Four others are in operation in 1910.
*g* Two others already authorized will open in September, 1910.
*h* "State agricultural high schools" are maintained in connection with 34 agricultural colleges. See Office Expt. Stas. Circ. 83, p. 21, May 27, 1909.

high schools for their respective communities, notably in New Hampshire and Vermont. The Alabama schools noted in the tables are the "congressional district agricultural schools." It is the plan of the state department of education soon to include

TABLE 2

SECONDARY SCHOOLS CLASSIFIED AS AGRICULTURAL HIGH SCHOOLS BY LAW, POPULAR REPUTATION, LOCAL CLAIMS, OR ON THE MERITS OF THEIR CURRICULUM, FOR THE YEAR 1908–9

| State | Number | |
|---|---|---|
| Alabama | 9 | (new county high school not included) |
| Arkansas | 0 | (see note "*e*" under Table 1) |
| California | 2 | |
| Georgia | 12 | |
| Louisiana | 9 | |
| Maryland | 1 | (now 2) |
| Massachusetts | 3 | |
| Michigan | 1 | |
| Minnesota | 1 | |
| Mississippi | 6 | |
| New York | 3 | |
| Oklahoma | 2 | (now 6) |
| Pennsylvania | 1 | |
| Virginia | 5 | (now 10) |
| Wisconsin | 5 | (7 by September, 1910) |
| Total | 60 | for 1908–9; including those since |
| located or authorized | 76 | |

a four-year course in agriculture in the new county high schools, 34 of which have already been organized. These do not appear in the tables.

Tables 1, 2, 3, and 4 include not only schools reporting directly, but those about which there is little doubt. Therefore, they are approximate rather than exact. The totals should probably be

TABLE 3

SPECIAL SCHOOLS CLASSIFIED ACCORDING TO THE SIZE OF THE GEOGRAPHICAL UNIT SERVED

| | |
|---|---|
| State high schools, not counting high-school departments located at the agricultural colleges | 8 |
| Special district schools of Arkansas (not yet started) | 4 |
| Congressional district schools of Alabama, Georgia, and Virginia | 29 |
| Judicial and other districts of Oklahoma | 6 |
| Total | 47 |
| County high schools of purely agricultural type | 13 |
| County high schools with agriculture and general courses | 29 |
| Total | 42 |

larger. Figures from Ohio township and village schools unreported would, no doubt, bring the total number of public high schools listed in Table 4 close to 500.

The special districts of Arkansas having from 17 to 20 counties each, are nearly twice the size of the congressional districts of the state. The judicial districts of Oklahoma, averaging about 15 counties each, are about the same size as the congressional districts but not conterminous with them. The "Panhandle" district has three counties.

About one-third of 35 district schools seem to be offering general high-school courses including classics.

The 8 county high schools noted offer two-year courses. Many of the others provide but two or three years of strictly high-school work.

Many of the county high schools with general courses are being organized, or reorganized to provide instruction in agriculture, and many others have doubtless not been reported.

In the following table are included high schools with courses in agriculture of one-half year or more. The geographical divisions and the classification according to population are those used in the reports of the United States Commissioner of Education. So far as known but one township has a population over 4,000, and that on account of the township high-school district containing the city of DeKalb, Ill., of about 10,000 population. The counties are kept separate, because the population is almost entirely rural, although they always run over 4,000 but scarcely ever include cities of over 4,000.

The classification of general high schools most serviceable for the purposes of this study is based on the political unit served by the school, which gives us the high schools of (a) county, (b) city, (c) village, (d) township, (e) special district, and (f) the quasi-public academies before mentioned, not drawing from a circumscribed area.

The county high school may not necessarily be, but usually is, situated at the county seat, and so is apt to have a high percentage of urban students, more so than in schools in villages of 2,000 or less. (See Table 10.) The township high school is often several miles from any village, but usually includes at least an unincorporated hamlet. The township high school often shares a building with a "consolidated" or "centralized" elementary school, and is then governed by the same school board and principal, but in some states it operates under a different board with separate taxing powers, as in Illinois. The special district high school may resemble any of the other types except the county high school. It may be supported by a city or village and the immediately surrounding territory; it may be composed of two or more contiguous country school districts, forming a sort of "part township high school"; or it may consist of parts of adjoining townships. The returns do not always indicate the nature of such special districts.

TABLE 4

PUBLIC HIGH SCHOOLS WITH COURSES IN AGRICULTURE

| Kind of School District | North Atlantic States | South Atlantic States | South Central States | North Central States | Western States | Total |
|---|---|---|---|---|---|---|
| County or part of...... | 0 | 2 | 29 | 7 | 4 | 42 |
| Township or parts of... | 4 | 1 | 1 | 89 | 0 | 95 |
| Villages less than 4,000. | *a*5 | *b*2 | *b*3 | *c-d*108 | 3 | *d*121 |
| Cities over 4,000....... | 1 | 1 | 3 | 24 | 1 | 30 |
| Total............ | 10 | 6 | 36 | 228 | 8 | 288 |

*a* In New York 35 villages and cities have agriculture in the training course but not in the regular four-year course.

*b* The table excludes a large number of village schools having a four-weeks' "review course."

*c* The table excludes a large number of villages doing no laboratory or demonstration work, and those schools professing to teach agriculture only incidentally.

*d* A large number of schools should be included in the table but returns necessary to classify them are not at hand; among these may be mentioned about 350 schools in the North Central States alone, including 222 in Nebraska not already counted. See page 30

The schools reporting may be grouped into two classes according to the type of instruction: those teaching agriculture as a distinct subject and those giving attention to it only as an incident in the teaching of the sciences usually given in high schools. This second class will receive its chief attention in the latter part of this chapter, with occasional references in appropriate sections of other chapters.

Over 75 per cent of the first class sent in more or less full reports following, in most cases, a preliminary postal card inquiry, while only 32 per cent of the second class so responded. This might seem to represent fairly well the relative importance attached to the work by the principals, and, judging from the check on their work by later questions, it probably represents equally well the degree of seriousness with which outsiders may regard the work done. The absence of experiments by the pupils or demonstrations by the teacher in the sciences given, or the use of any of the government bulletins and similar literature, seems to furnish a basis of fact for the contempt held by many for all so-called "incidental" or "correlated" agricultural work. The fact should not be overlooked, however, that not a few high schools listing agriculture as a separate subject for a half-year or longer are equally lacking in both the tools and the methods of instruction. It will also be seen from the notes on certain schools given in the next chapter, that a very high grade of agricultural instruction is sometimes given in connection with other science work.

The following tables show the distribution according to regions, type of schools, total enrollment in the various schools, enrollment in the agricultural classes, proportion of such classes to total enrollment, percentage from farm homes, and total number of such pupils. All the remaining tables in this chapter are based on returns received from the schools themselves.

Tables 5, 6, and 7 show the distribution according to regions, population, and number reporting more or less fully. The

TABLE 5

AGRICULTURE TAUGHT AS A SEPARATE STUDY

| Kind of School District | North Atlantic States | South Atlantic States | South Central States | North Central States | Western States | Total |
|---|---|---|---|---|---|---|
| Cities of 4,000 and over. | 0 | 1 | 6 | 19 | 1 | 27 |
| Cities of less than 4,000. | 4 | 4 | 1 | 146 | 3 | 158 |
| Counties (*a*) . . . . . . . . . . . | 0 | 1 | 3 | 6 | 1 | 11 |
| Total . . . . . . . . . . . | 4 | 6 | 10 | 171 | 5 | 196 |

*a* The counties are kept separate in Tables 5 and 6 because they always run over 4,000 but are essentially rural in population, and often include no towns with over 4,000 population.

geographical regions and rough population units are those used in the reports of the United States Commissioner of Education. The population is that of the city, village, township, or county supporting the school. A number of academies reported which

TABLE 6

AGRICULTURE TAUGHT INCIDENTALLY

| Kind of School District | North Atlantic States | South Atlantic States | South Central States | North Central States | Western States | Total |
|---|---|---|---|---|---|---|
| Cities of 4,000 and over. | 1 | 1 | 1 | 6 | 1 | 10 |
| Cities of less than 4,000. | 5 | 0 | 0 | 30 | 2 | 37 |
| Counties (*a*) . . . . . . . . . . . | 0 | 2 | 1 | 3 | 1 | 7 |
| Total . . . . . . . . . . . . | 6 | 3 | 2 | 39 | 4 | 54 |

*a* The counties are kept separate in Tables 5 and 6 because they always run over 4,000 but are essentially rural in population, and often include no towns with over 4,000 population.

are not included in these tables because of their not belonging to a definitely limited territory.

While most of the reports received are for the scholastic year 1907-8, a considerable number of the reports were made out for the year 1908-9, and came from schools in which the subject

TABLE 7

SCHOOLS REPORTING MORE OR LESS FULLY

| Kind of School District | North Atlantic States | South Atlantic States | South Central States | North Central States | Western States | Total | Per cent |
|---|---|---|---|---|---|---|---|
| Of those in Table 5 | 4 | 3 | 6 | 126 | 3 | 142 | 75 |
| Of those in Table 6 | 6 | 0 | 0 | 10 | 2 | 18 | 32 |
| Total . . . . . . . . | 10 | 3 | 6 | 136 | 5 | 160 | |

probably was not taught the previous year. A few other reports made out for the year 1908-9 are from schools in which agriculture was taught the previous year. A small number of schools included in Table 6 reported for the year 1908-9, having been addressed under the impression that they offered instruction in agriculture as a separate branch. The principals of all the other schools included in Table 2 were addressed early in 1908. Late in the same year names of over 100 schools reputed to be giving incidental instruction in agriculture were obtained, to which inquiries have not been sent. That agriculture is taught or not in all the schools included in the above tables has been verified by reports received from the superintendents, principals, or special instructors. If the ratio of schools teaching agriculture to those not teaching it, as shown by the returns from the list furnished for Ohio, were maintained by those schools which have not reported, about twenty would be added for that state for the year 1908-9. Official lists from certain other states have so shrunken when checked up by the returns received directly from the schools that not much is to be expected from those not replying. A few schools included in Table 6, and a number not included in either, have reported that they would offer distinct work along agricultural lines during the school year 1908-9. Fifty-seven high schools in Nebraska reported, some on the training class work in the eleventh and twelfth grades, some on the agriculture exclusive of the training class, and some on both. Twelve reported all of the pupils studying agriculture as being in the training classes. Lack of complete replies makes it impossible to differentiate accurately for those reporting only on the training classes. Information from State Superintendent E. C. Bishop shows that in 1909-10, there were in Nebraska 202 high schools teaching agriculture in the ninth grade, 25 in the tenth grade, 16 in the eleventh grade, and 24 in the twelfth grade, total 267. A few of these schools may and probably do teach agriculture as a general high-school subject, and in the eleventh or twelfth grades for training students only. The returns from Nebraska schools lead to the belief that these figures can be relied upon as substantially correct.

During the year 1909-10 some seventy local training classes in the State of New York are carrying on work in agriculture in connection with the nature-study. Table 1 includes none of the New York schools except the very few that have courses in agriculture, or teach it incidentally to the general student. The table also omits the numerous high schools in one state that offer a "four-weeks' course" to prepare for the county examination for third-grade certificate. Advices from educators of prominence who have visited nearly all of these schools, made it seem advisable to omit them from the enumeration. Their work is probably less profitable than the same time spent in well conducted laboratory work in botany or chemistry involving the principles underlying scientific agriculture. The official reports of several states credit a number of schools with classes in agriculture while the schools themselves have reported the

TABLE 8

DATA FROM 151 COMPARABLE SCHOOLS FURNISHING COMPLETE INFORMATION

| Kind of School District | Population served | Enrolled in high school | Enrolled in agriculture | From farm homes | No. of schools |
|---|---|---|---|---|---|
| Cities of 4,000 or more (*a*) | 118,000 | 2,367 | 478 | 455 | 13 |
| Cities and villages less than 4,000 | 114,775 | 5,529 | 1,657 | 2,014 | 83 |
| Townships (*b*) | 66,225 | 2,506 | 785 | 1,618 | 45 |
| Counties (*c*) | 161,200 | 1,954 | 308 | 1,234 | 10 |
| Total | 460,200 | 12,356 | 3,228 | 5,321 | 151 |

*a* Includes "special" or "consolidated" districts composed of a village and immediately surrounding territory, usually less than 6 square miles.
*b* Includes towns and cities of whatever size, when all are under the same township board and pay the same rate.
*c* Includes cities under the same conditions mentioned for townships under (*b*).

work as being only "incidental." Since they report little or no experimental work done in the sciences either by pupils or teacher, doubts naturally arise whether any of the science work is worthy of the name.

During the year 1908-9, 6 Michigan high schools had introduced four-year courses in agriculture, 10 more Minnesota high schools had established departments of agriculture, manual training, and home economics, and 34 county high schools were organized in Alabama which will be teaching agriculture within a year or two in order to receive recognition. By 1910, 7 parish (county) high schools in Louisiana had added agricultural departments, or were newly established with such departments, and the number of high schools adding departments of agriculture, manual training, and home economics in Virginia, had risen to nine.

Some reports omitted one item and some another, so do not appear in Table 8. The grand totals given below are for all the schools reporting on the several items without regard to whether they reported on every item or only part of them. The number of schools reporting on each item is indicated by the figures in parenthesis. A greater number of cases does not necessarily include all of those reporting on an item showing a smaller number of cases.

Grand totals for all the schools reporting on any of the above items: Population of districts (181), 543,950; enrolled in high school (188), 15,977; enrolled in agriculture (176), 3,726; from farm homes (164), 5,666; total schools, 188.

These totals include returns from normal training classes in Nebraska high schools that often have other students in agriculture who are not in the training class, and who in some of these cases were not reported.

For the year ending June 30, 1909, 335 high schools reported to the United States Department of Agriculture, a total enrollment of 54,700, with 9,500 in the agricultural classes.

It will be noticed from Table 8 that the 83 smaller cities and villages, having less population than the 13 larger ones, have two and one-third times as many pupils enrolled in the high schools, four times as many taking agriculture, and report almost three and one-half times as many pupils as being from farm homes. (See Table 10.) The remarkable fact is that the 13 larger cities should have even a fifth of their enrollment in the classes in agriculture, especially since Athens, Ga., with a population of 16,000 reports no pupils from farm homes, and Lake Charles, La., also with 16,000, reports but 4 per cent from farm homes. The two cities contain about one-fourth of the 445 pupils enrolled in the agricultural classes in the cities of 4,000 and over. However, the subject is a required one in these high schools, and most of their 103 students in agriculture are in the eighth grade, which is usually the first year of high school in the South, instead of the last grade of the elementary schools as is the case generally in the West and North. The counties are kept separate in the tabulations because they are apt to contain large elements both rural and urban. The high schools of the small towns of 2,000 or less, take on many of the characteristics of the township high schools, and for the purposes of this investigation show more similarity to them than to those of the cities above that figure. Table 9 is an attempt to display the data of the 151 comparable high schools, which are distributed in the table so as to reveal better their true nature. Table 10 seeks to show the relationships between the enrollment of the various types of school and the population of the supporting community. The numbers show the high-school enrollment for each 10,000 of population, and likewise the enrollment in the classes in agriculture and the number reported from farm homes. It should be mentioned that some latitude has been used regarding the last item. Under this head many principals reported children of parents living in town but owning and operating farms. Others in small villages report that "nearly every one gardens." Consequently the term "from farm homes" more truly means those having an immediate interest in agricultural occupations. The broken horizontal line in Table 10, separating the communities of less than 4,000 from those above, is intended to emphasize the difference in the upper and lower groups, and the similarity between the right and left groups.

While it seems natural to expect that the ratio of the items in the third and fourth columns should rise in the schools of the smaller communities, it is interesting, to say the least, to note the similar rise in the ratio of the enrollment of the high school itself. Whatever causes operate to lessen the ratio in the larger towns will also operate in the case of the counties and townships of the larger population, such as those tabulated, where a large share of the population is concentrated in the principal city. This is true of the one township having over 8,000 population, but of only one of the seven counties with over 8,000, and of two of the three counties with a population between 4,000 and 8,000. The remaining two are essentially rural. But this seems insufficient to explain the very low ratios in the case of all of the counties, a ratio almost as low as in cities of over 8,000 in regard to total enrollment, and lower in regard to enrollment in the agricultural classes. The low percentage of attendance may be due to the distance the pupils must travel if they do not board away from home. The low percentage of the agricultural classes would seem to indicate the determination of the pupils to get away from the farm.

TABLE 9

COMPARATIVE DATA FOR 151 SCHOOLS

| Civic Limit | No. of schools | Approximate population | Enrolled in high school | Enrolled in agriculture | From farm homes |
|---|---|---|---|---|---|
| Cities and villages: | | | | | |
| 8,000 and over | 4 | 65,000 | 758 | 193 | 69 |
| 6,000 to 8,000 | 5 | 34,000 | 892 | 175 | 153 |
| 4,000 to 6,000 | 4 | 19,000 | 717 | 110 | 233 |
| 2,000 to 4,000 | 26 | 61,200 | 2,475 | 595 | 831 |
| Under 2,000 | 57 | 53,575 | 3,054 | 1,062 | 1,183 |
| Total | 90 | 232,775 | 7,896 | 2,135 | 2,469 |
| Counties: | | | | | |
| 8,000 and over | *a*8 | 155,200 | 2,072 | 241 | 1,118 |
| 6,000 to 8,000 | 1 | 7,000 | 82 | 13 | 33 |
| 4,000 to 6,000 | 2 | 9,000 | 120 | 74 | 108 |
| Townships: | | | | | |
| 2,000 to 4,000 | 10 | 25,100 | 686 | 229 | 429 |
| Under 2,000 | 34 | 31,125 | 1,500 | 536 | 1,164 |
| Total | 55 | 227,425 | 4,460 | 1,093 | 2,852 |
| Total for all schools | 151 | 460,200 | 12,356 | 3,228 | 5,321 |

*a* One is a township containing the county seat and a state normal school.

These ratios would be more instructive if statistics of the attendance of boys and girls could be separated. The tables certainly furnish a strong argument for those who would strengthen the village and township high schools along agricultural lines. The argument may be more valid for the richer and more populous states than for others with a public high-school system less fully developed.

The tables show, as one might naturally suppose, that most of the country children found in high schools are in those of the villages and townships. They also show the extent of the relationship between the schools and this constituency.

Data in the World's Almanac, for 1910, compiled by the statistician of the Bureau of Education, show for the total population of United States, an enrollment in public secondary schools of 91 for every 10,000 of population.

These ratios for high schools outside of cities having a population of 8,000 and over do not differ greatly from those obtained by Dr. Snyder in his study on rural schools.[8] His ratios and

TABLE 10

RATIOS OF ENROLLMENT PER 10,000 OF POPULATION

| Population | Cities and villages | | | | Counties and townships | | | |
|---|---|---|---|---|---|---|---|---|
| | No. of schools | Enrolled in high schools | Enrolled in agriculture | Reported from farm homes | No. of schools | Enrolled in high schools | Enrolled in agriculture | Reported from farm homes |
| 8,000 and over... | 4 | 115 | 30 | 11 | 8 | 133 | 16 | 73 |
| 6,000 to 8,000.... | 5 | 262 | 51 | 45 | 1 | 117 | 19 | 47 |
| 4,000 to 6,000.... | 4 | 372 | 57 | 122 | 2 | 133 | 82 | 12 |
| 2,000 to 4,000.... | 26 | 405 | 97 | 134 | 10 | 273 | 91 | 171 |
| Under 2,000..... | 57 | 570 | 198 | 221 | 34 | 481 | 172 | 374 |

these for the three states, Missouri, Nebraska, and Ohio, are shown below, his appearing in the first column.

Missouri, 332 schools, 2.52 per cent; 35 schools, 3.86.[9]
Nebraska, 365 schools, 5.29 per cent; 47 schools, 4.51.
Ohio, 782 schools, 5.06 per cent; 41 schools, 3.78.

A high percentage of pupils enrolled in agricultural classes may mean (1) that the work is required instead of elective,

[8] E. R. Snyder, The Legal Status of Rural High Schools in the United States, p. 136.

[9] Most of the school districts included in my returns have less than 2,000 population, namely, 25 in Missouri, 18 in Nebraska, and 37 in Ohio. Dr. Snyder's ratios for cities of 8,000 and over run much higher, being 4.22 per cent, 7.16 per cent and 4.75 per cent respectively for the three states in the order already named. I have returns from no cities of this size in these states, but it will be observed that the cities I do cite above this population, mostly in the South, give a much lower percentage, as do also the county units. But in some of the counties there are other high schools drawing from the same population.

(2) that work is offered in more than one year of the course, (3) that the work is given in alternate years to two classes at once, and (4) that the curriculum of the school is only two or three years long or, what amounts to the same thing, that the school has been organized but one, two, or three years. (1) About three-fifths of the schools reporting on this point require the subject. All of the 60 or more Nebraska high schools maintaining training classes require agricultural work of stu-

TABLE 11

RATIO OF ENROLLMENT IN AGRICULTURE TO TOTAL ENROLLMENT

| Per cent | Cities of 4,000 and over | Counties and townships of 4,000 and over | Villages of 2,000 to 4,000 | Townships of 2,000 to 4,000 | Villages under 2,000 | Townships under 2,000 | Total cases |
|---|---|---|---|---|---|---|---|
| 91–100 | .. | 1 | 1 | .. | 3 | 4 | 9 |
| 81–90 | .. | .. | .. | .. | .. | .. | 0 |
| 71–80 | .. | .. | .. | 1 | 2 | 1 | 4 |
| 61–70 | .. | .. | 1 | .. | 6 | 1 | 8 |
| 51–60 | .. | .. | .. | 1 | 6 | .. | 7 |
| 41–50 | 2 | .. | 3 | .. | 9 | 4 | 18 |
| 31–40 | 4 | .. | 2 | .. | 14 | 8 | 28 |
| 21–30 | 3 | 3 | 7 | 1 | 18 | 12 | 44 |
| 11–20 | 3 | 2 | 10 | 4 | 7 | 4 | 30 |
| 0–10 | 4 | 5 | 7 | 3 | 4 | 3 | 26 |
| Total | 16 | 11 | 31 | 10 | 69 | 37 | 174 |

dents in those classes, but make it elective for others. As is natural, the smaller the school, such as the new township schools of Ohio, the more certain the subject is to be required of all students. It is usually placed in the first or second years. (See Chapter IV.) (2) Not more than 12 schools included in the full returns offer agricultural courses in more than one year. These are found in all groups and in all sizes of schools. (3) No data are at hand on this point, but such an arrangement of giving the work in alternate years has been reported several times. (4) No accurate data are available, but the fact has been mentioned several times in reports from township high schools that "this is our first year," or second year, as the case might be.

Table 11 shows that in over three-fourths of the schools, the ratio of the enrollment in classes in agriculture to the total high-school enrollment lies between 11 per cent and 50 per cent, with a strong "mode" between 21 per cent and 30 per cent, almost one-third of the schools being within this range.

Table 12

Ratio of Students from Farm Homes to Total Enrollment

| Per cent | Cities of 4,000 and over | Counties and townships of 4,000 and over | Villages of 2,000 to 4,000 | Townships of 2,000 to 4,000 | Villages under 2,000 | Townships under 2,000 | Total cases |
|---|---|---|---|---|---|---|---|
| 91–100 | .. | 1 | 1 | 3 | 1 | 14 | 20 |
| 81–90 | .. | .. | .. | 3 | 1 | 7 | 11 |
| 71–80 | .. | 3 | .. | 2 | 3 | 6 | 14 |
| 61–70 | .. | 1 | 1 | .. | 2 | 3 | 7 |
| 51–60 | .. | 1 | 2 | 1 | 6 | .. | 10 |
| 41–50 | 1 | 3 | 2 | .. | 12 | 5 | 23 |
| 31–40 | 3 | 1 | 8 | .. | 14 | 1 | 27 |
| 21–30 | 1 | 1 | 4 | .. | 13 | 1 | 20 |
| 11–20 | 7 | .. | 4 | 2 | 4 | .. | 17 |
| 0–10 | 5 | .. | 4 | .. | 1 | .. | 10 |
| Total | 17 | 11 | 26 | 11 | 57 | 37 | 159 |

Table 12 indicates that over one-half of the schools have between 11 per cent and 50 per cent of their total enrollment made up of pupils from farm homes. The strongly bimodal distribution in Table 12, it will be readily seen, is due to the numerous township high schools that are being so rapidly organized in Ohio, and that are so rapidly introducing agriculture into their course of study. From September, 1906, to November, 1908, the number of township high schools reporting to Professor Graham of Ohio State University that they were doing work in agriculture increased from 10 to 70, and of these the cases verified in this study increased from 9 to 37. Many of these township districts contain no incorporated village.

The size of these high schools throughout the entire country that are introducing agricultural work is indicated by Table 13, which gives all verified cases whose enrollment could be learned from any source, both for schools teaching the subject separately and for those professing to do agricultural work incidentally.

TABLE 13

ENROLLMENT OF THE HIGH SCHOOLS REPORTING

| Enrollment | Teaching agriculture separately | Teaching agriculture incidentally |
|---|---|---|
| 25 or less | 16 | 5 |
| Between 26 and 50 | 49 | 15 |
| Between 51 and 75 | 29 | 4 |
| Between 76 and 100 | 25 | 1 |
| Between 101 and 125 | 16 | 1 |
| Between 126 and 150 | 13 | 4 |
| Between 151 and 175 | 2 | 0 |
| Between 176 and 200 | 3 | 1 |
| Between 201 and 250 | 6 | 2 |
| Between 251 and 300 | 4 | 7 |
| Between 301 and 350 | 2 | 1 |
| Between 350 and 400 | 2 | 0 |
| Over 400 | 2 | 1 |

## STATISTICS OF SCHOOLS TEACHING AGRICULTURE INCIDENTALLY

Seventeen schools reporting agriculture as taught in connection with other sciences, and which give comparable data, are in districts with a population of 53,855, and have an enrollment of 1,767, with about 600, or over one-third, from the country.

TABLE 14

RATIO OF STUDENTS FROM FARM HOMES TO TOTAL ENROLLMENT

| Number of schools reporting | Enrollment of pupils | Per cent of pupils from farm homes |
|---|---|---|
| 5 | 1,047 | 0–25 |
| 4 | 213 | 26–50 |
| 4 | 233 | 51–75 |
| 4 | 197 | 76–100 |

Thus the five cities with an average of 13 per cent from farm homes have 32,700 inhabitants, over half the total population of the districts tributory to these schools, and 1,047, or more than half of the total enrollment. In striking contrast with this, the remaining 12 schools, in communities with less than

TABLE 15

SCIENCE WORK REPORTED

| Science | Number of times reported |
|---|---|
| Botany | 30 |
| Chemistry | 10 |
| Physics | 7 |
| Physical geography | 26 |
| Zoology | 10 |

20,000 inhabitants, enroll 720 students in their high schools, of whom approximately 460, or 64 per cent, are estimated to be from country homes. But even in the larger cities reporting, the number of families represented who depend for their livelihood upon the calling of agriculture is probably as large pro-

TABLE 16

NUMBER OF SCIENCES REPORTED

| Number of schools reporting | Science reported |
|---|---|
| 17 | 1 |
| 10 | 2 |
| 77 | 3 |
| 5 | 4 |
| 1 | 5 |

portionally as the number depending on any other single vocation. This being so, it deserves at least an incidental treatment in the sciences as truly as do steel making and photography in chemistry, or electric wiring and telephony in physics. Almost any one may have the opportunity to do something, as an amateur, with vegetable and flower gardening, and many do so, while any one cannot do so in most of the other trades, as most people are effectually prevented from casual participation in them.

Of the 58 schools reporting agriculture taught in connection with other sciences, 37 have specified the sciences involved, 8 or 10 were not asked, and the remaining schools did not reply to the question.

TABLE 17

TIME GIVEN EACH SCIENCE

| Botany | | Chemistry | | Physics | | Physical Geography | | Zoology | |
|---|---|---|---|---|---|---|---|---|---|
| No. of schools | Minutes | No. of schools | Minutes | No. of schools | Minutes | No. of schools | Minutes | No. of schools | Minutes |
| 1 | 1,680 | 1 | 1,680 | .. | ..... | .. | ..... | 1 | 1,680 |
| .. | ..... | .. | ..... | .. | ..... | 1 | 1,800 | .. | ..... |
| .. | ..... | .. | ..... | .. | ..... | 1 | 2,400 | .. | ..... |
| 4 | 3,000 | .. | ..... | .. | ..... | .. | ..... | .. | ..... |
| 1 | 3,600 | .. | ..... | .. | ..... | 2 | 3,600 | .. | ..... |
| 2 | 4,000 | .. | ..... | .. | ..... | 2 | 4,000 | 1 | 4,000 |
| .. | ..... | 1 | 4,800 | 1 | 4,800 | .. | ..... | .. | ..... |
| .. | ..... | .. | ..... | .. | ..... | 1 | 5,040 | 1 | 5,040 |
| 1 | 7,200 | .. | ..... | .. | ..... | 1 | 7,200 | 1 | 7,200 |
| .. | ..... | 1 | 8,000 | .. | ..... | 1 | 8,000 | .. | ..... |
| 1 | 8,640 | .. | ..... | .. | ..... | .. | ..... | .. | ..... |
| .. | ..... | .. | ..... | 1 | 10,080 | .. | ..... | .. | ..... |
| 1 | 10,480 | .. | ..... | .. | ..... | .. | ..... | .. | ..... |
| .. | ..... | 1 | 14,400 | .. | ..... | .. | ..... | .. | ..... |
| .. | ..... | .. | ..... | 1 | 15,120 | .. | ..... | .. | ..... |
| .. | ..... | 1 | 16,200 | 1 | 16,200 | .. | ..... | .. | ..... |

In many of these schools botany and physical geography were the only sciences taught, sometimes only one being given.

A convenient unit of high-school work is an eighteen-weeks' course, five times a week, forty minutes a day, which totals 3,600 minutes. If the study be carried through thirty-six weeks, we should have 7,200 minutes. The number of weeks reported in the several schools varied from twelve to forty; the number of minutes from 120 to 540 a week; and the time given to work "other than recitation," from 60 to 270 minutes a week. The largest number of minutes of laboratory work is found in schools giving the most time to the subject.

Out of 33 schools reporting on experimental work, 12 stated that no experimental work was done in science, 9 reported that "some" or "very little" was done, leaving 12 answering "yes" without qualification. Of these 12, 4 specified the work as being principally demonstrations by the teacher, and 2 that it was chiefly individual work done by the pupils. Three of the 12 gave less than 25 per cent of the time to work other than recitations, 4 gave from 25 per cent to 50 per cent and 5 gave 50 per cent or more.

Ten schools included in Table 17 and only 15 of the entire 37 schools reporting agriculture as taught incidentally, stated that they made use of agricultural bulletins, reference books, etc. As the texts specified for the various subjects contain little or no reference to the agricultural application of the principles of the several sciences, it will be apparent that many of the claims that agriculture is taught in connection with these sciences rests upon a doubtful basis, and that it is often impossible to estimate the value of such work without a personal visit, which has been made in some cases.

Table 17 gives the time devoted to the several sciences obtained by multiplying the minutes per week by the number of weeks for the subject. Fifteen schools furnished the data necessary for this table.

# CHAPTER III

## SOME TYPICAL HIGH SCHOOLS TEACHING AGRICULTURE

A number of schools scattered between the North Atlantic coast and the Missouri River were visited in May and June, 1908. These included fairly representative examples of county, township, and village high schools. The largest was in a county seat, a number were several miles from a railroad, and a few were some distance from the nearest village. The agricultural courses in these schools were partly of the half-year type, partly of one year, and in four cases were planned for the entire four years, though not always in full operation. In most cases there was opportunity to see the classes, and to inspect their work and the school equipment. There was also some slight opportunity to learn the sentiment of the community.

### High Schools Teaching Agriculture One Year or Less

Very early on this itinerary a group of schools was observed whose work in agriculture was similar principally in being carried on for but half the year. In two schools the agricultural work was taken up during the winter of the second year, preceded by a study of the plant, recognized as a part of the more formal botany, and followed by a few weeks spent in identifying spring flowers. In one school the same plan was followed in the third year. In another the agriculture of the last half of the first year was preceded by physical geography, while still another school placed it in the first half of the same year with physical geography following. Under the prevailing arrangement one might suspect that there was not the amount of outdoor work that might have been done had the course extended through the entire year. The various forms of home work will be noted under the treatment of the respective schools.

The schools described of this group were visited consecutively, though not in the exact order in which they are described, and before any of the other schools, except the one at Waterford, Pa. The work first described is in the schools of three villages in central and western Ohio, and of one in southern Michigan, having a population of 1,000 or less. The next three schools treated are distinctly rural high schools, two in southwestern Ohio, and one in the northwestern part of the state, with approximately the same sized constituencies as the village schools. The last two schools discussed are in larger villages of about 3,000, one in northern Michigan, and one in western Iowa.

### *Grove City (Ohio) High School*

The first place visited in the state was Grove City, Franklin County. It is a typical Ohio village about 20 miles south of Columbus. It is essentially rural in character, with no important manufacturing industry, and has a population of about 1,000. The total enrollment of the school was 55, one-half of whom came from farm homes. The class in agriculture numbered 33, with about the same proportion of farmers' children. Only a few of the high-school pupils teach school after leaving. One or two of the senior class of three expected to attend college. There were two high-school teachers besides the superintendent, who devoted about one-half of his time to the grades.

The instructor used no text but worked by the topical method, following a recently published text of distinctly high-school grade as a guide, and assigning references in the bulletins with which the school was abundantly supplied. The bulletins most used were those relating to special plants studied in the work, such as cabbages, pansies, and tomatoes. In this school, the first half year's work was devoted to physical geography, and the last half to agriculture. No special attempt was made to introduce agricultural topics into the geography work. As yet most of the experimental work that has been tried, both in agriculture and in botany, which comes the second year, has been carried on out of class and at home.

A feature giving excellent results has been the excursions

made by teacher and class. Each trip was made the basis of the next recitation. The students were also held responsible on the monthly review for the knowledge gained on these trips and in the subsequent discussions. In this way the trips were kept from degenerating into picnics.

A visit was made to an orchard that was just being set out, where the members of the class planted some of the trees. As the instructor did not profess to be an expert in the nursery business, the attitude of both teacher and pupils was that of learning together. Another visit was made to an orchard of a successful plum grower, where observations were made on his methods of caring for the ground, the distance at which he planted his trees, and other details. It was noted, for instance, that he had some trees planted but ten feet apart, which were not thriving so well as were the others. The only class work done in grafting has been in the nature of exemplification with sticks having the bark on and not with real roots and scions, and without reference to the proper season. The following is a brief account of the planting done at home, and is here given, not because of any variety of startling features, but as an example of the possibilities.

The class was consulted as to the materials to be used. The girls wished mostly to try pansies. Many of the boys and some of the girls wished to work with tomatoes, others preferred cabbages. They got the seeds and all started the work together about the middle of March, and brought their plants to school for an exhibit during the week of May 4th to 8th. The instructor indicated for the benefit of the class what plants he considered to be the " winners." All students made oral reports, those having poor success as well as the others. The work in English was in the hands of the high-school assistant and no attempt was made to correlate it with the agriculture. The pupils were required to read the bulletins and other available literature before starting the seeds to germinating, so as to inform themselves on the conditions of soil, shade, moisture, temperature, and other factors. The class had studied the general topic of soils before starting this germination. One boy made the mistake of planting the seeds of all these varieties in one box and found, of course, that all did not do equally well, as they required different treatment, a fact that was very apparent in the exhibition. The pansies fared the worst, suffering from the amount of sunlight given the others. This served as a "control" experiment, although the instructor had made no provision for one. A garden trowel was given as a prize to the winner of this competition.

The class also took up the subject of bees and read John Burroughs' "Birds, Bees, and Sharp Eyes." The instructor, who had had some experience with bees, brought a hive to school. The boys took measurements from it and made hives at home. This was not, however, a part of the manual training that was carried on as school work, although the principal said that the construction of bee-hives during the winter might well have been assigned as one of the pieces required. He instructed the pupils how to get bees. He told how he had caught two wild swarms the previous summer, and suggested that probably they could persuade a beeman to give them a late swarm to nurse over the winter. Bee culture bulletins were then assigned for study.

One boy took as a special topic the treatment of potatoes for scab. Having no ground of his own, he received permission to experiment on a plat about to be planted by a neighbor who had been rather unsuccessful with his potatoes. The result of the experiment convinced the owner of the efficacy of the treatment. There was, however, no control experiment that I could learn of. Another boy so convinced his neighbors of the benefit of spraying potatoes that they joined together to buy a good sprayer.

The instructor instituted an agricultural club that met through the following summer, and reported a very successful series of meetings. Each member agreed to experiment on some crop and report the results. The instructor also gave each member a dozen strawberry plants for trial.

### *New Holland (Ohio) High School*

The environment and influence at New Holland, in Pickaway County, are, in many respects, similar to those in the village just described. The places are of about the same size but the latter had by far the more modern plant. Although the New Holland school is in a special district of six square miles, the enrollment in the high school was somewhat smaller, being forty, and the percentage of pupils from farm homes was rather less than half the percentage at Grove City. In each case the teaching force consisted of two instructors in addition to the superintendent.

The community is engaged in general farming and ships nearly all the produce. Comparatively little live stock is raised, but the land is kept in condition by a three or four-year crop rotation including clover. A canning factory offers inducements to the farmers to raise sweet corn, tomatoes, lima beans, peas, etc., for canning.

The smaller size of the agricultural class at New Holland, numbering ten, was due to the fact that the study was taught in the third year instead of the first. It is preceded by sixteen weeks of botany in the fall. The subject is taught by the superintendent, who has shown a great deal of interest in it. He has made a collection of bulletins larger than is usually found in schools devoting more time to the subject, and has worked out a complete and handy system of filing, and has had the pamphlets carefully card-catalogued by the students. He has also issued a topical index of the bulletins of the state experiment station and of the college of agriculture. Corn growing being a very important industry, considerable attention is paid to testing the viability of the seed. About forty ears were tested besides considerable shelled corn. The ears came in lots of from one to three each. All tests were on corn actually to be used for seed.

On Arbor Day the class planted an apple tree which the members had top-grafted after preparing their own grafting wax and cloth binding-strips. The tree had been grafted in the winter with apple and quince scions, and had been kept in the cool cellar until Arbor Day. Four other small trees, apple, cherry, plum, and quince, with plenty of superfluous branches, were brought in and trimmed up as part of the Arbor Day exercises, according to directions for pruning in the Farmers' Bulletin on this topic, after which the trees were thrown away. The exercises followed the program sent out by School Commissioner Jones. Although no class work has been done on spraying trees or treating seeds, one boy has informed himself on the subject and has done considerable of it. Samples of fertilizers were obtained from the local agents of seedmen. Some of these were in neat cases of cloth-covered paste-board. The formulas were studied by the class, although no chemistry was given in the school outside of demonstrations before the class in physiology.

The school arranged a series of public lectures on agricultural and nature-study subjects by professors of the State University and others. This school then held the trophy offered by the College of Agriculture of the State University for the best exhibit of seed corn. This trophy is offered to students under eighteen years of age in high schools teaching agriculture. The names of all winners are engraved on the cup.

### *West Milton (Ohio) High School*

While West Milton, in Miami County, differs little in population from the two villages already mentioned, its high school enrollment was considerably larger, being 80, of whom 40 were from farm homes, and almost all from homes where gardens are made. The high-school instruction was carried on by three teachers, agriculture being taught by the superintendent, who was himself reared on a farm and now owns one. His aim has been to make the course a sort of general introductory science course. The compound microscopes and stereopticon are used in this course as well as in the botany, which is closely linked with it. Many specimens were brought into class to illustrate scale, blight, and other fruit-tree diseases to which the instructor has given considerable attention, but no spraying was tried. Visits were also made to stock farms. The attitude of the community is well shown by the fact that 52 of the 53 first and second year students selected agriculture in place of English history, which it was supposed the girls would choose.

Six were later persuaded to enter the history class. Over half of these 46 were children of parents living on farms or owning them.

### *South Lyon (Mich.) High School*

At this point may be given an account of the work at South Lyon, Mich., a village in the southern part of the state, with but a few hundred more inhabitants than the Ohio villages already described, and similar to them in many ways. Stock raising, dairying, and trucking are the principal industries besides general farming. The village had a rather large "retired farmer" element, without representation in the public schools, who opposed attempts to better the rather antiquated and utterly inadequate school facilities. In contrast to this was the progressive body of younger farmers who were favorable to the agricultural instruction and whose children composed the 55 per cent non-resident portion of the high school. Most of the remainder of the seventy pupils enrolled were closely identified with agricultural interests. The majority of the sixteen members of the agricultural class were girls of the third year. Nearly all of the twelve agricultural students the previous year were boys. Several of these were planning to attend the Michigan Agricultural College on graduation. One boy living in town expressed himself as having been "converted to the profession" through his interest in the work of the class. There was little "elimination" apparent. The second year class has usually been larger than the first, and all of the entering class of two years before were still in school. The tuition charge of $12 represents approximately the per capita cost of instruction. The agricultural work was in charge of the superintendent, who was also principal of the high school. He very wisely did not allow the school machinery to interfere with the effectiveness of his work, and took part of a day for field work whenever it seemed worth while. Visits were made to a creamery, an agricultural implement store, and to stock farms, a very good one being within a mile of the school. The instructor's teaching methods in science and mathematics seemed better than those one is likely to find in most small high schools. In botany the plant analysis had been reduced to fifteen plants and emphasis was laid upon the reasons for plant and flower structures and upon their possible origins. The botany class had examined the actual grafting done on the tree by the agricultural class, and did "sample" grafting in the class room similar to that done at Grove City, O. The previous year each member of the class made two grafts on an old apple tree in the school yard, bringing their own scions. The reason given for the current class not doing this work was that the members were mostly girls! The class of the year before had also made an extensive test of corn, each student testing five ears to be used on his own place. Most farmers of this vicinity select their seed from the field; but few take it from the crib. One or two samples of the latter kind showed very poorly in the test, and formed a good check or control on the other lots. The soil experiments were given as a part of the agriculture and not of the physical geography, which came later. The extensive truck gardening for the Detroit market and the varied morainic topography of the region make the study of soils and drainage a very important one. An interview with some ditch diggers near the town suggested some very interesting problems along this line. The superintendent made field trips, and work with the Geological Survey sheets constitutes an important part of this work.

Twice a week the class in agriculture debated some topic, going to bulletins and to farmers as well for their arguments, and giving the authorities quoted. It is often girls against boys. with more girls than boys in order to overcome any handicap caused by the superior general information the boys might have. These debates were not a feature of the regular work in English, which was in charge of another instructor. Many of the farmers quoted by the debaters are of superior intelligence, several being graduates of the Michigan Agricultural College. A number of these have given lectures before the school, and have also taken great pains to indicate the "points" in the live stock inspected and to pass judgment on the scoring done by the pupils.

### *Wayne Township High School, Lee's Creek, Ohio*

The Wayne Township centralized school, in Clinton County, Ohio, is a good example of another type of schools. It differs from those already described in being several miles from the nearest railroad and incorporated village. Although the nearby hamlet of Lee's Creek numbers scarcely more than 200 inhabitants, the school serves a township with an area of 36 square miles and a population of 1,200. Naturally all of the 27 pupils of the high-school department were from farm homes. The agriculture is taught in alternate years, the last class numbering 12. Eighteen weeks, of 200 minutes each, were given the subject, extending over the entire year. It is a second-year study, preceded by a half-year each of physical geography and botany. The text used is one of the popular books written for the elementary grades. The superintendent found that it proved too easy, as the boys already knew most that it contained. The bulletins on corn and on weeds, issued by the state experiment station, were the ones most used as references. Agriculture was placed in the course because of the unanimous demand made by the patrons. The spirit of the community is well illustrated by the question of one of the most conservative of the farmers, himself at first opposed to consolidation, who asked the superintendent why the school did not try to be up-to-date and introduce domestic science. Plans are being definitely made for this and for manual training as well. The basement is large and airy, and well adapted for such work. One part of it, not occupied by the heating plant, was used as a lunch room and physical laboratory. The building was erected three years ago at a cost, including equipment, of $17,500, and has a complete water, lighting, and heating system. Professor Graham pronounces it the finest centralized school building in the state.

General farming and stock raising form the principal industry. There is little dairying or gardening. The farmers feed all they raise. One makes the boast that he has never sold a dollar's worth of grain from his place. Of course these conditions largely determine the nature of the class work. The most important feature of the experimental work was that done at home, especially with corn. The students had the advantage of being able to observe the ten-acre plat of a local seed-corn breeder, and to note his method of detasseling the stalks. Much rivalry arose in a boys' corn contest, in which one of the boys managed to excel the exhibits of this gentleman's son, according to the judgment of the class and the agriculturist present from Washington. Ordinary seed was used the first year the experiment was tried, while the second year the seed used was from ears selected from the boys' own plats and tested before planting. As many as fifty or seventy-five ears were tested, and from one to a dozen ears in a lot were brought to school for the trial. Corn testing is generally practiced by the farmers of the district, due quite largely, the superintendent thinks, to the influence of the school work. Owing to a bad season three years ago, it was necessary to test the seed in order to be sure of a stand. One year the boys planted seed sent out from the State University, which did not yield as well as the home grown seed planted by their fathers. The fathers in this case were able to use the failure to point the lesson that home grown seed is better than that shipped in, especially from a different latitude, as this was, having been raised fifty miles northeast. The general plan followed by the boys was to plant 288 hills, enough for two shocks, in the corner of the father's field, and to leave the two outside rows uncultivated. As there was much rain early in the season, the boys reported a marked difference in the growth under the two kinds of treatment. This furnished the application of the principle of capillarity that had been studied experimentally in the laboratory. The test of the corn growing experiment seems to have been entirely qualitative, as measured by the selected ears; whereas it might have been measured quantitatively, either absolutely or comparatively, by weighing the yield and computing from that the rate per acre.

About twenty-five experiments in all were performed to illustrate osmosis, capillarity, etc. Some of the simpler chemical phenomena were demonstrated by the students, and others by the instructor. The school did not then possess even the modest apparatus it has since acquired. Fifty dollars a year is being appropriated for science equipment. There are two teachers in the high-school department besides the superintendent, who gives it about half his time.

*Beaver Creek Township High School, Alpha, Ohio*

The Beaver Creek Township High School, at Alpha, Greene County, differs from the one just described, in existing separately from the graded schools, which are but partially consolidated. This accounts for the fact that the pupils are not transported in wagons as in Wayne Township, but drive their own horses. The thirty animals used are housed in comfortable stables, built on one corner of the school grounds at a cost of $1,000. In the high school building provision is made for home economics and manual training in the basement. The equipment of the latter consists of twenty benches and tools. There is no indebtedness nor will there be any for the addition that may have to be built to accommodate the manual training. The present school levy is six mills on a valuation of $1,800,000.

The campus consists of five acres or more of partially cleared forest with a few apple trees. The high school is of eighteen years' standing. The teaching staff consists of two teachers besides the superintendent, who gives half his time to the elementary schools. The 52 pupils are practically all from farm homes, as there are but two small hamlets of a hundred or less each in the township. The agricultural class numbered nine. The succeeding class was expected to enroll about 20 students. While the class of the current year had all of the botany except plant analysis in the fall, the next class is to take up the agriculture in the first half of the second year. It will be preceded as heretofore by a half-year of physical geography in the first year. None of this work is elective.

Most of the experiments in the text were performed, many by the pupils individually. The text used is one of the more recent ones of distinctly high-school grade. The pupils went into the fields and selected the corn to be used for testing. The corn was judged on score cards ruled by the pupils for the purpose. This work was not competitive. Individual study was made of the vitality of seeds, using corn, oats, wheat, etc.

The school possessed a supply of seeds four years old kept to show the influence of age. Each student was assigned one insect and one bird to investigate thoroughly from the literature and from field observations, and was required to hand in his report at the end of the year. There were no school gardens, or home gardens under school supervision or direction. The children from the hamlets seem to take more interest in the work than those from the farms.

The principal interest of the community centers around raising cattle, hogs, and sheep. Shorthorns are the favorite cattle. There are no dairy herds in the strict sense, although many farmers have from five to ten milch cows and ship the milk to Dayton by the traction line. Very few of the graduates of the high school teach after leaving, at least not in the township; for none are employed without a year of normal training. From $45 to $65 a month is paid in the elementary grades, which attracts teachers with the necessary training from the nearby city.

### *Riley Township High School, Pandora, Ohio*

The schools of Riley Township and Pandora, in Putnam County, exist by virtue of legal provisions enabling villages and townships to maintain jointly a system of high and elementary schools. In this case neither township nor village could well support a high school alone, as the population of the entire district is only 800. One hundred were enrolled in the high school, two-thirds from farm homes. The agricultural class numbered 35, 20 boys and 15 girls, and was taught by the superintendent. He had a man for principal of the high school department and a lady assistant. Agriculture was followed by physical geography in the first year, each continuing half the year. Botany was placed in the second year.

The school is in a rich agricultural country. Corn is the principal crop. Most of the field produce is fed to cattle and hogs. There is considerable dairying, the milk being taken by a creamery. This of course has a milk tester, but the school has none. All the corn testing was done at home by the boys of the class. They tested all the seed to be used on their farms the following spring. One had 54 ears, or about half a bushel. Five kernels from each ear were tested. Of the 270 kernels, 260 germinated, making over 96 per cent. Another tested 100 ears, getting a result of 95 per cent. He rejected 2 ears but did not enter the percentage of those ears in his record. A third tested 49 ears. Some of the reports that were inspected showed great care and a truly scientific attitude toward the work.

The girls were supposed to do a little work in plant propagation with geranium slips. Owing to the clumsy work done by the girls individually in previous years, the instructor was doing most of the work for the members of the class. Considerable work had been done with insects. Samples of scale were brought in. Various insects that had been collected were displayed in glass mounts made by the janitor. These were glass boxes 3 by 5 inches, with glass tops and bottoms spaced 3/8 inch apart, and separated by four wooden sides and corners.

It may not be amiss to speak here of certain points that could not well be noted under the individual schools. These seven schools are in communities of nearly the same population and having very similar interests. None of the schools enrolls more than 100 pupils. The teaching force ranges from two to three instructors, including the superintendent, who must give some time to the supervision of grade work. The instructors in agriculture with one exception are men under middle age. Nor does age in the one case seem to have any inhibitive effect on enthusiasm or a progressive spirit, for this gentleman has attended several summer sessions at the University of Chicago and has since earned his degree. But two of the seven held degrees from colleges requiring four years for graduation. The college work of the other five ranged from two years or less to three and one-half years. One of these was a normal school graduate. Several had been pursuing summer courses and were planning to continue them. Two of the seven grew up on farms. Another, whose father moved to town in the future superintendent's early boyhood "to give the boys a chance," has since acquired considerable practical experience in farm management. In all cases there seemed to be essential agreement between authorities and patrons as to the desirability of having agricultural work in the curriculum, and agreement among the superintendents as to its educational value as compared with other subjects they were teaching. Many will be willing to admit that the judgment of men of some experience in teaching unrelated subjects may be of value on this point, perhaps equal to that of specialists. For the latter, even with their more thorough scholarship in restricted lines (and perhaps on account of it), must often form their judgments on a priori grounds or from a one-sided experience.

### *St. Louis (Mich.) High School*

St. Louis, Michigan, a village of 3,000 inhabitants, is in Gratiot County, in the north central part of the state. The high-school enrollment is between 70 and 80. The superintendent teaches the class in agriculture and botany, numbering between 25 and 30. This work is in the second year and is followed by zoology in the third year and chemistry in the fourth. The printed course of study calls for a one-year course in "orchard, garden, and field crops," to follow the elementary course just mentioned, but the superintendent had not yet put it into operation as he wished the second-year course to be placed on a firmer basis. In such a course certain topics would be expanded that are now treated only briefly in the elementary course. While the work is required in only one course leading to graduation, called the agricultural course, it has been elected for several years by a large number of pupils not in this course.

The school has a good equipment for its size, including two compound and thirty-five dissecting microscopes, and an amount of chemical apparatus sufficient for individual work by the small class. The science classes have ample room at their disposal. Many of the usual indoor experiments are carried on with home-made apparatus.

Some dairying is carried on in the neighborhood, and the class had visited a creamery, although no work in milk testing had been attempted. The class was taken to a diseased peach orchard, where the students pruned and trimmed out the diseased and dead limbs of all the trees except one, which was left as a "control." The class was told that the untreated tree would probably infect the nearest trees. Orchards in the county are threatened with extinction from insect pests, according to the superintendent, with, as he tersely expressed it, "a cat to every bird." The curculio and other injurious insects are discussed and identified, and talks are given on the appropriate method of treatment for each.

Demonstrations of the different methods of grafting were made in school, after which each student at the proper season grafted a limb on a tree at home. At the close of the school year, the limb was sawed off and brought to school as part of the record of the year's work. Many seeds were tested but not corn, except in the general experiments, such as those showing the effect of drainage in tin cans and others of like nature.

One of the most important local industries is sugar beet culture and the manufacture of the sugar. Notwithstanding the fact that the beet is preeminently the "money-maker" of the region, not more than one acre in ten or twelve is devoted to its culture. In line with this industry, beet-sugar analysis is a prominent feature of the year's work in chemistry. The students are taught the practical use of the polariscope. Several young men with no other preparation than this have taken positions as assistant chemists in the local plants at a better salary than is usually commanded in village positions. The class as a whole, so far as learned, had not visited the sugar factories, but most of the students had done so individually.

A majority of the board and many of the patrons favor the further development of the agricultural line of instruction. It is significant that so many adults bought copies of the school text-book that the students were not able to supply themselves until the bookstores had received a second and third supply.

### *Sac City (Iowa) High School*

The constituency and enrollment of the high school at Sac City, Iowa, is somewhat larger than those already described. The work in agriculture has been conducted along different lines and presents many points of interest. The population of the town is approximately 2,800, and of the district, 3,600. The district, which is 4 miles square, lies mostly in one township but extends into three others. About 20 per cent of the enrollment of 136 comes from the country homes, and about half of the 34 or 35 country children elect agriculture. This work has been carried on for six years. During the first year 28 boys and 30 girls took up the study and 56 finished the course. Fifty were in the class the second year and 52 the third. The lowest number registering for agriculture in any one year was 30. The study is an extra elective and does not find a place in the program of the regular school day but is counted toward graduation. The recitations run through the winter months while the practical work may continue to the end of the year. Students are enrolled from all years of the high school, while recently a few capable boys of the seventh and eighth grades were allowed to enter the class. They were competent to make the observations, report on home experiments, and could follow the botanical but not the chemical part of the work. Seniors and juniors predominated in the earlier years, but recently the major part of the class has been from grades below the junior year.

Though another man has charge of the other science work, agriculture is taught by the superintendent. He became interested in the scientific study of agriculture because he owned a farm worth $110 an acre that was not yielding an adequate return on that capitalization. For his own guidance he uses King's "Physics of Agriculture" and his book on "The Soil," and Osterhout's "Experiments with Plants." The students have no text, but use Office of Experiment Stations Bulletin 186, "Plant Production," by Crosby, and refer to other bulletins. The school subscribes for two farm journals.

While raising corn and feeding stock is the chief industry of the community, dairying also holds a place of importance. Sac County and Ida County adjoining it are said to raise one-third of the world's supply of popcorn, the annual yield being valued at from $35 to $60 per acre. The community has shown a great interest in the agriculture taught in the school. It has enlivened the interest in the farmers' institutes, and has caused the general adoption among the farmers of the practice of selecting and testing seed-corn and of measuring the milk product of the dairy herds.

The school possesses two Babcock testers, each of four bottles capacity. One is without a shell and the other is of the covered type but runs very hard. Milk was first studied the second winter the course was given. Different classes have tested the milk of no less than 200 cows, distributed probably among 40 or 50 herds. The farmers finally brought in more samples of milk than the class had time to handle, especially after the class had published a report on work done with a cow belonging to one of the patrons. He weighed the milk one day, estimated the total product as closely as possible to fit the approximate period of lactation, and had the class test samples of the milk. The result was later published in the local press. Then a ten-year-old son of the Methodist minister bought a cow for $45. This animal was used as a miniature experiment station for the class during the year. The class figured out the rations to be fed. The boy kept all the data for the use of the class, such as rent ot pasture and stable, cost of feed and labor, and receipts from the sale of milk. The milking period was estimated at 300 days. The milk was weighed and tested. The amount of butter product was calculated at the rate of seven-eighth of a pound of butter-fat to a pound of butter. The cost of the feed for the year was calculated, and the profit was then computed from this data. Later several farmers bought testers. The brown Swiss is the favorite dairy breed in this section. Most of the farmers now breed their own dairy cattle. Breeds have not been studied as such by the class nor have attempts been made at judging stock, although the various parts of the animal carcass have been studied from charts.

The use of a piece of land at the edge of town was secured for the class the first year, so that every member had a plat, but the results were interfered with by outsiders disturbing the beds. Nevertheless the children brought in quite complete reports the following autumn. Many of those living in the country who had to come some distance took better care of their plats than did others living near by. Since then the village children have had plats at home in which they grow the usual garden vegetables, while the country children mostly put their plats into seed corn. The village children use their own taste largely and put in early garden truck and potatoes. Much of the work is experimental, testing different methods. Thus with potatoes, certain students divided their plats into four parts. In one part were planted quarters of the potato using the seed end, in another part quarters from the stem end, in both cases from potatoes that had been exposed to the sun for about four days; while in the third and fourth parts were planted the quarters from potatoes just taken out of the dark cellar bin. The results were carefully written up showing the yield of large and medium potatoes, and the unmarketable part of the crop. One girl tried the experiment of raising two crops of potatoes, using an early variety. She succeeded in getting tubers from the second crop weighing from four to six ounces.

The boys from the farms have directed most of their efforts toward breeding corn, bringing in notes of their work in the fall the same as the village children. Their corn has been judged by a representative from the Iowa State Agricultural College, at Ames. An experiment in selecting and breeding was started in 1904 by two brothers who were then seniors, which has since been continued after they went to college by a younger brother in the high school. As a result of the experiment, they have produced a perfectly tipped ear, even to the one central tip kernel. They concluded, however, that it did not pay, as the process has "pinched the ears shorter." They had kept the ears of previous years for comparison. These two brothers had become so interested in scientific agriculture that they decided to go to the college at Ames and fit themselves to become farm superintendents. Other influences diverted them to a denominational school, where they have just graduated. Most of the graduates from the high school go to sectarian colleges and very few to the agricultural college. The superintendent believes that five of the last senior class of twenty would not have entered high school from the rural schools but for the favorable impression made upon their fathers by the work in agriculture.

## High Schools Teaching Agriculture Three of Four Years

This section is devoted to the description of the work of four schools in which the course in agriculture is planned to extend through the last three years of the course or through the entire four years, and in which the curriculum is well under way. The school in the town of Petersham, Mass., and also the John Swaney School of Magnolia Township in Putnam County, Illinois,[1] had been in existence two years at the time this study was made, and minister to districts distinctly rural. The school at Waterford,[1] Pa., is maintained jointly by the borough and township and has a history of almost a century. Its agricultural department was established in 1904. The Guthrie County High School, at Panora, Iowa, is thirty years old and has been offering courses in agriculture for several years, long enough to have all scheduled courses in full operation.

One sees then that these schools, widely distributed as they are, may appropriately be taken to represent the most progressive efforts to work out through the high school this important problem in districts widely separate and different in their interests.

### *Petersham (Mass.) High School*

The Town of Petersham has a population of about 900. The high-school enrollment was 18 the first year and 37 the second. Forty-five is the estimate made for the third year's enrollment. The courses leading to graduation are the college preparatory, the academic, and a short course of three years. Agriculture or domestic science and floriculture are required in all years of the last two courses, replacing the required Latin of the first course. The total number of hours required will be 320 hours of agriculture for the boys, and for the girls, 240 hours of domestic science and 80 of floriculture. These are in addition to the required work in botany, zoology, physiology, physics and chemistry. The curriculum is shown on page 63. At present

[1] The John Swaney School was visited also by County Superintendent O. J. Kern, and the Waterford High School by D. J. Crosby of the U. S. Department of Agriculture, whose comments on the schools and the class work they saw form part of the report of the standing committee of the National Education Association on Industrial Education in Schools for Rural Communities, for 1908.

the work of instruction in agriculture is carried on by the principal, who also has charge of the discipline of the elementary grades, a lady assistant who gives all her time to the high school, and a grade teacher who handles the domestic science. The classes were combined so that studies not consecutive are taught only on alternate years. The principal at the time the school was visited, Mr. Edw. H. Scott, B. S., is a graduate of the agricultural college at Amherst, and of the Worcester high school, where he received his instruction in manual training. Of the five graduates the second year, the two boys planned to go to the agricultural college, two of the girls to normal school, while the third girl expected to take care of a home of her own.

The handsome stone and brick school building was the gift of James W. Brooks. The campus of eight acres furnishes the ground for recreation and sports, as well as for the practical work in agriculture. Already half an acre has been given over to gardening purposes, and a small orchard has been set out. Owing to certain opposition the work started out the first year under the guise of geology, and included an elementary study of soils and soil physics, a scheme that the rocky condition of New England soil made plausible enough. Later in the year plant propagation was taken up with practical work. At this time the "greenhouse" was a modest glass lean-to built against a barn across the street. Then a systematic course was carried on for some weeks on the study of potatoes, pruning, grafting, and budding. The pruning was done on a lot of trees already growing on the campus. The grafting was in the way of demonstration. During the second year, the course included a study of fertilizers in class, with testing, greenhouse culture of lettuce, and cucumbers, the growth of plants for setting out, and floriculture.

This work was made possible by the erection during the previous summer of a greenhouse just south of the school and connecting with it through the basement. The house is 18 by 30 feet, with a stone substructure, and with an independent heating plant. Had it been necessary to pay for the labor and material, the cost would probably have been between $800 and $1,000. The cucumbers were started in April, and began to produce by

June 1. They found a ready market at 5 cents apiece. The blossoms were pollenized by a hive of bees, which was placed outside and close to an opening in the greenhouse. The young tomato plants were sorted into three groups according to stockiness and apparent hardiness, and notes were made by the class on the difference in their ability to withstand the transplanting, which was very marked. An exhibit of vegetables and flowers raised in the greenhouse was held late in the spring. Some of the boys conducted an experiment in treating for scab the potatoes used for seed by a blacksmith of the village. The effect was so marked that the experiment gained considerable local fame and gave the agricultural work an immediate standing in the community. It was the first thing told the principal on his return the succeeding fall. Most of the soil experiments have been conducted in the greenhouse. The school has a six-bottle Babcock milk tester of the covered type.

The elementary physics in the first year is a prerequisite for all other science work. It is purely demonstrational, of the "phenomena" type, and presented with a somewhat limited set of apparatus, which serves, however, to illustrate most of the commonly used principles of physics. The seniors and juniors formed a class in chemistry running through two terms, and followed by botany in the spring term. The instruction in chemistry was of the same nature as in the physics, except the individual work done by two post-graduate students who performed individual experiments at the demonstration table. The class in bookkeeping used Masury's text with problems dealing largely with the things that make up the bulk of farm accounts, and not with the merchandise common to most texts in the subject.

The community is engaged in dairying, raising hay and oats for feed, and to some extent in the cultivation of potatoes and corn "by hand." Probably from three to four acres out of five are in timber or in "sprout lots," that is, young timber coming up on cut-over land. Half of the remainder is under actual cultivation and the other half is in blue grass or "herd grass" (timothy). Gardening under glass is carried on to a large extent a few miles north of Petersham.

## CURRICULUM OF THE PETERSHAM HIGH SCHOOL

[*a*, *b*, and *c*, placed after a subject, indicates first, second, and third term, respectively.]

| College preparatory course | | Academic course | | Short course | |
|---|---|---|---|---|---|
| Subjects | Hours per week | Subjects | Hours per week | Subjects | Hours per week |
| | | *First Year* | | | |
| Algebra | 4 | Algebra | 4 | Algebra | 4 |
| English | 2 | English | 2 | English | 2 |
| Ancient history | 3 | Ancient history | 3 | Ancient history | 3 |
| Elementary physics | 3 | Elementary physics | 2 | Elementary physics | 3 |
| Spelling 5 | | Book-keeping | 3 | Book-keeping | 2 |
| Drawing 2 | | Spelling 5 | | Spelling 5 | |
| Music 2 | | Drawing 2 | | Drawing 2 | |
| Rhetoricals 1 | 3 | Music 2 | | Music 2 | |
| Latin | 4 | Rhetoricals 1 | 3 | Rhetoricals 1 | 3 |
| | | Agriculture 2 (boys) | | Agriculture 2 (boys) | |
| | | Domestic science 1 (girls) | 2 | Domestic science 2 (girls) | 2 |
| Total | 19 | Total | 19 | Total | 19 |
| | | *Second Year* | | | |
| Geometry | 4 | Geometry | 4 | Geometry | 4 |
| Elementary chemistry | 3 | Elementary chemistry | 3 | Elementary chemistry | 3 |
| Spelling, music, etc. | 3 | Spelling, music, etc. | 3 | Spelling, music, etc. | 3 |
| English (*a*) | | English (a) | | American history (*a*) | |
| English history (*b*, *c*) | 3 | English history (*b*, *c*) | 3 | Civics (*b*, *c*) | 3 |
| French or German | 4 | French or German | 4 | Physiology (*a*) | |
| Latin | 4 | Manual training | 2 | Botany (*b*, *c*) | 2 |
| | | Agriculture 2 (boys) | | Manual training | 1 |
| | | Domestic science 1 (girls) | 2 | Agriculture 2 (boys) } Domestic science 1 | 2 |
| Total | 21 | Total | 21 | Total | 18 |
| | | *Third Year* | | | |
| English | 2 | English | 2 | English | 6 |
| Spelling, etc | 3 | Spelling, etc | 3 | Spelling, etc | 3 |
| French or German | 4 | French or German | 4 | Chemistry (*a*) | |
| Biology (*a*) | | Biology (*a*) | | Trigonometry and Surveying (*b*, *c*) | 4 |
| Botany (*b*, *c*) | 3 | Botany (*b*, *c*) | 3 | Astronomy (*a*) | |
| Physiology (*a*) | | Physiology (*a*) | | Zoology (*b*) | 3 |
| Physics (*b*, *c*) | 2 | Physics (*b*, *c*) | 2 | Geology (*c*) | |
| Latin | 4 | American history (*a*) | | Manual training | 1 |
| | | Civics (*b*, *c*) | 3 | Agriculture 2 (boys) | |
| | | Manual training | 1 | Domestic science 1 (girls) | 2 |
| | | Agriculture 2 (boys) } Domestic science 1 (girls) | 2 | | |
| Total | 18 | Total | 20 | Total | 19 |
| | | *Fourth Year* | | | |
| English | 4 | English | 4 | | |
| French or German | 3 | French or German | 3 | | |
| Spelling, etc | 3 | Spelling, etc | 3 | | |
| English (college requirements) | 2 | Astronomy (*a*) | | | |
| Latin | 4 | Trigonometry and Surveying (*b*, *c*) | 4 | | |
| Chemistry (*a*) | | Chemistry (*a*) | | | |
| Mathematical reviews (*b*, *c*) | 4 | Zoology (*b*) | | | |
| | | Geology (*c*) | 4 | | |
| | | Agriculture (boys) } Floriculture (girls) | 2 | | |
| Total | 20 | Total | 20 | | |

The "Center" claims to be the highest cultivated place in the state and has a deeper deposit of boulder till and less rock outcrop than most of the hills in the vicinity. Many engage in farming in the summer and work during the winter in mills not far from the town limits. In a hat factory at North Dana, six miles west, a number of women get as high as $20 to $25 a week, and bright girls often become sufficiently proficient in one or two years to make fair wages. One farm was pointed out that had just been purchased by a married couple both of whom work in the mills in winter.

Summer residents form no small part of the village population and take an interest in the schools. All the high-school pupils of the town and a few non-residents attend here, but two or three school districts have not yet consolidated. The country children are brought to the building in school wagons.

### *John Swaney School, McNabb, Ill.*

The John Swaney School, in north-central Illinois, furnishes a good example of what may be done in an intelligent rural community, independent of any urban influences, and working for a school adapted to its own peculiar needs. It lies in the open country three miles from McNabb the nearest post-office. This is merely a country crossroads with two or three stores and a station on one of the short coal roads in that mining region of the state. The school is on a beautiful campus of twenty-four acres presented by one of the patrons and the principal benefactor, Mr. John Swaney. He is a well preserved man of eighty-five living a short distance from the school. He takes a keen interest in the school and makes frequent visits to it. The campus includes land under cultivation and a tract of rolling open timber with a large variety of trees, only one or two of which were felled to make room for the building. On it stands a building used as a district school before the consolidation but now used as a dwelling by the teachers, a grange hall, a roomy horse shed, and the new school building. The latter is a three-story building, on the second floor of which there is an assembly room accommodating 200 or 300 persons and one or two recitation rooms. On the first floor are the rooms for the elementary grades, the domestic, natural, and physical science laboratories, and the principal's office, while in the basement are the lunch and manual training rooms, and the heating and gas plants. The latter supplies the laboratories and furnishes light.

The school was opened for work January, 1907. When the curriculum is in full operation agriculture will be offered part or all of each of the four years, a total of six semesters' work in all.

CURRICULUM OF THE JOHN SWANEY SCHOOL

| First semester | Second semester |
|---|---|
| *First Year* | |
| English I<br>Algebra<br>Physiology<br>Agronomy I or Latin I<br>Household science or manual training | English I<br>Algebra<br>Physical geography<br>Horticulture or Latin I<br>Household science or manual training |
| *Second Year* | |
| English II<br>Algebra, 10 weeks<br>Geometry, 10 weeks<br>Zoology<br>Ancient history<br>Drawing | English II<br>Geometry<br>Botany<br>Ancient history, 10 weeks<br>Animal husbandry, 10 weeks or Household science, 10 weeks<br>Music |
| *Third Year* | |
| English III<br>Chemistry<br>Agronomy II or Latin II or household science<br>English history | English III<br>Chemistry<br>Animal husbandry or Latin II or household science<br>English history |
| *Fourth Year* | |
| English IV<br>Physics<br>Agronomy III or household science<br>American history | English IV<br>Physics<br>Bookkeeping, 10 weeks<br>Arithmetic, 10 weeks<br>Civics |

As will be seen from the course of study, two semesters of agriculture and two semesters of manual training are required of all the boys, with an option of four semesters of agriculture in place of the Latin. The girls are required to take four semesters of some phase of household economics, and have an option of two more in place of Latin.

Physiology and physical geography are required in the first year of all students, botany and zoology in the second, chemistry in the third, and physics in the fourth, making a total of four years of science required besides the agriculture and home economics. The laboratory sciences (all except those of the first year) have three double periods and two recitation periods given them. This is the usual time allotment in most of the better high schools in the Central States. The agricultural work consists of a semester each of agronomy (I), horticulture (I), animal husbandry (II), agronomy (III), animal husbandry (III), and agronomy (IV). The Roman numeral indicates the year in which it occurs. Latin is offered theoretically in the first and third years, allowing the first and second-year pupils to be combined in one class. Of the 25 in these first two grades, 7 take Latin and 18 agronomy and horticulture. The following year the then first and second year students were scheduled to take ancient history. The current second and third year classes were scheduled to take English history in place of the botany and zoology, while the elective agronomy and animal husbandry will be offered to them.

For one study the agronomy class picked weed seeds and labelled the phials in which they were placed. The pupils had practice in sorting out the red clover, alfalfa, and other seeds of the clover group, which were mixed, likewise radish and rape seeds, timothy and grass seeds. The object of the work was to gain an ability to detect adulteration in bought seeds. Corn judging and testing was included as a part of the work.

The class in horticulture planted some trees, and did some grafting in an orchard nearby. In response to a question about three-fourths of the class stated that they had fruit trees at home, and most of them thought their fruit trees had received different treatment because of the work done in the class.

Upon request, the principal gave the following question in a written test. "What did you do at home as the result of the work done in horticulture this year?" Some of the more pretentious answers are here given.

"I have pruned several peach trees and shade trees, and have shown the hired men how to cut the limbs on the big orchard trees, which were too much for me to do alone. A one year peach tree, which I pruned to a whip, is one mass of leaves and twigs. The wounds are all healing nicely."

"I have pruned one peach tree and have watched the marked growth of a young soft maple that was headed in severely. It has very thick foliage, and some of the growths are from 9 to 12 inches long. These growths were all made since the middle of March."

"The young orchard now has a cover crop upon it. I have also helped make the grafting wax used in budding and grafting; from which good results have been had so far."

Another reported doing some grafting on his uncle's orchards. All four scions on one of two apple trees grafted were doing well. After thinning and pruning another small tree that was shrubby and worthless, he grafted eight scions of a different variety, all of which were growing well.

One student pruned grapevines, apple, peach, pear, plum, and cherry trees at home, and did some spraying.

One student reported that he gave the grapevines more pruning than they had before, pruned nearly all the orchard, used a spraying machine *bought this spring,* and "burnt the cut-off limbs instead of leaving them in a pile for a breeding place for fungus and insects." Bailey's "Principles of Fruit Growing" and his "Pruning Book" are the books most used.

The following references were read as a part of the regular work of the class in agronomy:

Shaw's "Weeds and How to Eradicate Them."
Farmers' Bulletin 97, Alfalfa Hay for Hogs.
Farmers' Bulletin 194, Alfalfa Seed.
Farmers' Bulletin 260, Seed of Red Clover and its Impurities.
Farmers' Bulletin 73, Kafir Corn.
Farmers' Bulletin 92, Planting and Replanting Corn.
Farmers' Bulletin 225, Testing of Seed Oats.
Farmers' Bulletin 97, The Soy Bean as a Feeding Stuff.

Farmers' Bulletin 105, Spontaneous Combustion of Hay.

Farmers' Bulletin 144, Rotation of Crops.

U. S. Dept. Agr., Office of the Secretary, Circular 18, The Adulteration of Red Clover Seed.

Illinois Agr. Expt. Sta. Circular 81, The Potato and the Selection of Seed.

U. S. Dept. Agr. Yearbook, 1901. Agricultural Seeds.

Topical references on the following subjects were also read as a foundation for the work in testing seeds for purity and viability: Extent and importance of the trade in agricultural seeds; centers of production; classes of agricultural seeds; cereals; corn; buckwheat; oats; rice; clover and other leguminous forage plants; grading and sampling clover seed; how clover seed is bought and sold; red clover seed; mammoth, alsike, and white clover; crimson clover; red clover and sweet clover; alfalfa; cowpeas and soy beans; canada field peas and hairy vetch; velvet bean and beggar weed; timothy; meadow fesque; orchard grass; Kentucky blue grass; red top, and beardless brome grass.

The type of work done in zoology is of interest as showing how a pure science may be made to minister to the cause of agricultural education. A mere mention of the text and the time spent in laboratory work would indicate only that the work met the requirements of the better colleges. Some distinctive features are shown by the following abstracts of reports taken from the note books of several students. The first is a condensed account of six field trips written by one of the girls. The others are random paragraphs from the reports of other students giving in more detail some of the observations made on the same field trips. The abstracts have been subjected to the least possible editing.

*Zoology Reports*

(1) Field trip, September 4. Purpose was to catch insects for study. Insects found were grasshoppers, potato bugs, caterpillar, and a spider. Field trip, September 6. To look for insects on trees. Found tent caterpillars and gall flies. Field trip, September 9. To watch tent caterpillars work. Also gathered some insects, as worms on elm leaves, pyramidal gall, sour-

bug, cox-comb gall, measure worm, cicada, grasshopper, katydid, black bug, butterfly, and spiders. Field trip, September 11. To gather insects. Found black moths, mosquitoes, green plant louse, larva of lady-bird beetle, leaf hopper, squash bug, larva of squash-vine borer. Field trip, October 2. To look for insects. Found cottony maple scale, plant louse, ants. Field trip, October 7. To look for insects in a corn field. Found caterpillars, corn-stalk borer, green plant lice, ants, corn-root aphis.

(2) Field trip, September 4. On this trip we also found three species of the old-fashioned potato-bug. It is also known by the name Spanish fly or blister bug. These bugs were working on potato and raspberry vines and grasses. One species was a plain black bug (*sic*) about three-quarters of an inch long. Another was about the same length, but it was black with a white stripe down the middle of its back and white stripes around its wings. The last one had dark brown stripes running lengthwise on its back. . . .

(3) On September 9 we went out on the campus and watched some tent caterpillars at work. We found a colony of eight worms on elm leaves. They were black with light stripes on them. There were six or seven clusters of brown hairs on each segment. . . . We found pyramidal galls on some elm leaves but there were no insects in them. They were small and oblong in shape. We examined the leaves of an elm tree and found the coxcomb gall. The encasement, which was empty, was dark brown and its upper edge was ragged and rough. . . .

(4) Field trip, September 13. We then went to a truck patch just across the road from the campus. We examined the stem of the pumpkin and in it we found several squash-vine borers. They are very injurious to the plant and often destroy it entirely. They are of a whitish color. One of the passage-ways of the borer was measured and it was found that it had bored over 3 feet, which destroyed the vine. The adult form of the borer is a moth but is not injurious, only that it lays eggs that hatch into one of the great pests of the pumpkin patch. Field trip, October 7. The zoology class went to a corn field north of the campus and examined some of the cornstalks. In them we found small brownish worms, called the cornstalk borer, which is the larva of a moth. In one stalk, we found punctures in every joint, in the first seven joints from the ground, two or three in every joint. We found another small stalk with no ears on it being worked on by the borer. The class thought that the borer was getting ready for the pupa stage, as they

did not seem to be working much. We found some green plant lice on a corn leaf, some with wings. A small stalk of corn that had an ant hill at the base, was pulled up by the roots and it was found that all the dirt had been worked away from them. The corn-root aphis were thick on the roots. They were a greenish color. We shook them and some ants off on the ground. The ants began carrying them away immediately. . . .

Certainly such work as indicated in these students' notes will not suffer by comparison with that of high-school classes spending their time on planarian worms, campanularion hydroids, and holothurians, all of which the writer has seen studied 1,000 miles from the sea coast. The eighth grade of a prominent eastern school system during the past year began the study of a series of animal types with the amoeba.

The principal of this interesting school, Mr. McNeil C. James,[3] is a graduate of Illinois Normal University, at Normal, and had, at the time of my visit, partly completed the course in agriculture at the University of Illinois. He later returned to the University and graduated with the class of 1909. He had prepared himself more especially along the line of agronomy. His successor, a graduate of the Northern Illinois Normal School, at DeKalb, took courses in animal husbandry at the summer session of the University last year, so as to be able to present that work to the two classes who had already taken the agronomy and horticulture. The assistant principal has studied at the University of Chicago and has had special training in domestic science at Teachers College, New York. By introducing a certain amount of departmental teaching in the grades, the principal has been able, with a small teaching force to give his school some of the advantages of larger schools with teachers especially prepared along certain lines.

Besides a fair amount of general apparatus, the school possessed special one-inch glass tubing for soil tests, two-inch metallic tubes with sealed and with perforated bottoms, larger galvanized iron vessels holding definite amounts of earth, and other equipment for soil investigations. An orchard of twenty-

---

[3] Mr. James now has charge of agricultural work in the Valley City (N. D.) State Normal School.

four young trees was planned, laid out, and set out by the class in horticulture. An experiment station of several acres is conducted on the campus by Mr. Griffith, a local corn breeder, under the direction of Professor Hopkins, of the College of Agriculture. On account of the extremely wet spring, the class in agronomy did not attempt to do any outdoor planting.

The consolidated district comprises fourteen sections of land and has a population of about 350, smaller than that of any other district visited. The assessed valuation is $164,000, which represents, under the taxation laws of Illinois, one-fifth the supposed real value. As all over the state, the board is allowed by the so-called Juul law to levy 2½ per cent for building purposes, and a like amount for maintenance. The total enrollment of the high school last year was 32. Several times this number are in the grades. Those living at a distance are transported in school wagons.

### *Guthrie County High School, Panora, Iowa*

Although Iowa has had on its statutes for many years a law providing for the establishment and maintenance of high schools by counties, the only one instituted so far under these provisions is the Guthrie County High School, at Panora. This school is finishing its thirtieth year and has outgrown two buildings. The classes in science and agriculture at present use most of the second of these, a two-story brick structure.

The village of Panora, with a population of over 1,000, maintains a school system of nine grades and promotes its graduates into the second year of the county high school. Two other villages in the county, with populations of 1,600 and 1,400 respectively have four-year high schools, while several villages of about 500 inhabitants each have one-year or two-year high schools. The total enrollment in the county high school for the year 1907-8 was 206; the number in the senior class was 30. Panora furnished 45 students, about 20 more than its quota. Each township is allowed free tuition for students up to a certain percentage of its population and must pay tuition for all over that number. The expense, which is between $6,000 and $7,000, is borne by the entire county. Guthrie County has a population of about 18,600. Five of the seven instructors held college degrees. One of the remaining two, Mr. J. I. Thompson, the instructor of agriculture, has since completed his senior year's work in the Iowa State Agricultural College.

CURRICULUM OF THE GUTHRIE COUNTY HIGH SCHOOL

[*a* and *b*, placed after a subject, indicate first and second semester, respectively.]

| Latin course | Science course | Agricultural course |
|---|---|---|
| | *First Year—all courses* | |
| Arithmetic, Algebra, American history, Civics, English, Latin, Physiology. | | |
| | *Second Year* | |
| Algebra<br>English<br>Ancient history (*a*)<br>Mediaeval and Modern history (*b*)<br>Latin | Algebra<br>English<br>Ancient history (*a*)<br>Mediaeval and Modern history (*b*)<br>Physical geography (*a*)<br>Botany and Bookkeeping (*b*) | Algebra<br>English<br>Ancient history (*a*)<br>Mediaeval and Modern history (*b*)<br>Animal husbandry<br>Agronomy<br>Bookkeeping |
| | *Third Yoar* | |
| Algebra (*a*)<br>Plane geometry (*b*)<br>German<br>Physics<br>Latin<br>Arithmetic (*a*) | Algebra (*a*)<br>Plane geometry (*b*)<br>German<br>Physics<br>Geology<br>Arithmetic (*a*)<br>Zoology | Algebra (*a*)<br>Plane geometry (*b*)<br>German<br>Physics<br>Geology<br>Farm mechanics |
| | *Fourth Year* | |
| Plane geometry (*a*)<br>Solid geometry (*b*)<br>German<br>Political economy (*a*)<br>English (*b*)<br>Latin | Plane geometry (*a*)<br>Solid geometry (*b*)<br>German<br>Political economy (*a*)<br>English (*b*)<br>Chemistry | Plane geometry (*a*)<br>Solid geometry (*b*)<br>German<br>Political economy (*a*)<br>English (*b*)<br>Chemistry<br>Soils (*a*)<br>Horticulture (*b*) |

About 50 students were enrolled in the three classes in the agricultural course. Three courses are offered—a Latin course, a science course, and an agricultural course. It will be seen from the appended curriculum that physics is the only science in the Latin course, and that three years of science in the scientific course replace the Latin, and that the agricultural studies replace in part the Latin and science of the other courses, and in part are added to the various sciences, so that it is not a "snap" course.

Since stock raising is the principal industry of the county, most interest centers around animal husbandry and the related subjects of corn growing and animal feeding. For this reason all such work is taught in the earliest year of the agricultural work. It has been found that this arrangement best enables the school to justify the work in the eyes of the patrons and appeals to the liveliest interest of the country pupils. Only boys are admitted to the class in animal husbandry, which shuts out girls from the agricultural course, unless allowed to make substitutions. As the principal of the school expressed it, "The farmers scoffed the first year, were silent the second, and approved of the work the third," for it had won their respect.

The following description of the courses in agriculture will indicate the scope of the work and the methods followed.

*Stock judging.*—Score card practice supplemented by lectures on the origin and history of the various breeds of live stock. Craige's work on stock judging is used as a basis of the study, and on specified days of each week the classes visit the stock farms of the surrounding country. The animals are studied from the standpoint of the farmer and professional stockman. The fall months are devoted to cattle, both beef and dairy types receiving attention. Horses and swine occupy the spring months.[4]

Two days a week are given to lectures and one to score-card practice and judging throughout the year. During the fall term the other two days are given to lectures on live stock management. Notes of all lectures are written up by the boys.

*Agronomy.*—The study of corn and other grains occupies those months of the school year when it is impossible to do field work. On account of its prominence, corn receives the greater attention. The ideas that Professor Holden has been emphasizing for the past few years are systematized and the principles of seed selection and seed testing put into actual use.

This work occupies two days a week during three of the spring months. Two-thirds of the time is devoted to the scoring and placing of corn, the remainder to making germinating tests, largely on samples of seed to be used in spring planting

[4] Guthrie County High School Bulletin, 1907–1908.

sent in by farmers. Holden's "A B C of Corn Culture" is the text.

*Farm mechanics.*—This subject is given two days a week throughout the third year. The students study printed sheets, prepared by the instructor on the development and mechanism of farm machinery. Considerable time is devoted to setting up, by the students themselves, various machines furnished free of charge by the manufacturers of farm implements and machinery. The school has had the use during the year of gasoline engines, mowing machines, corn harvesters, manure spreaders, plows, and other implements. At the time of the author's visit the students had just finished setting up for delivery a binder that had been sold. This was the first year the course had been given, and no little astonishment was caused by the possibility, as was learned by experience, of a failure to make a passing grade in the subject when the mechanical principles involved in the machines were not sufficiently grasped. In answer to a question, the principal ventured the opinion that possibly the course might be combined with the physics now taught the agricultural students in the same year with advantage to both studies.

*Chemistry and soil physics.*—All students in the science and agriculture courses take chemistry the first semester of the fourth year. The students of agriculture continue the subject the remainder of seven months, during which time it becomes differentiated as soil physics. In addition to the usual chemical apparatus the school possesses for this special work a drying oven and several glass and galvanized iron tubes for the study of moisture percolation, soil capillarity, etc. Williams' text is used in chemistry, and King's "Soil Physics" in the latter part of the work. Stevenson's "Laboratory Guide for Soil Experiments" is the basis of the practicums, which take up one-half of the time. All the experiments are performed by the students. Five periods a week are given to the study.

The chemical composition of soils, plant food, sources of fertility, texture, moisture with the causes affecting its movement and conservation, and similar topics are discussed and investigated in the laboratory, each student gathering and testing his own samples.

*Horticulture.*—The spring months of the senior year are devoted to this subject. Bailey's "The Nursery" and Goff's "Principles of Plant Culture" are followed, embracing the most important principles of fruit growing, grafting, cross breeding, and allied questions.

Only a small amount of laboratory work was done in this course, that being mostly in grafting and budding. These students had all taken botany in their second year. As future classes in horticulture will not have had the botany (see printed curriculum) the work must be modified to include the botanical material necessary to understand the scientific principles underlying horticultural practice. No garden work has been done in connection with the school work. Nor has any close correlation been worked out between such sciences as physics, geology, or physical geography, and the agriculture. That is, no agricultural content has consciously been put into those subjects.

The high-school board is composed of seven members, the county superintendent being president. The principal, H. E. Ilsley, M. S., has had considerable training in science and is making an effort to strengthen the science work as much as possible. The science work is in charge of the assistant principal, C. L. Cose, B. S. Much attention has been paid to the humanities as shown by the well selected school library of 2,000 volumes. The attitude of those in charge on the question of education for citizenship is shown, not only by the inclusion of political economy, history, and modern foreign language in the agriculture course, but also by the solicitude shown for the farmers' children—that none should be deprived of the best culture of our civilization, nor be made thus to feel any class distinction. A casual inspection of the course here printed will show how earnest an effort has been made to include those elements of a general course that are supposed to give a broad outlook, and to make the course at the same time truly practical from the agricultural standpoint. The large proportion (90 per cent) of pupils from families living on farms or especially interested in them shows the importance of such a view of the curriculum.

### *Waterford (Pa.) Township High School*

In rather marked contrast with the last two schools, with plenty of modern equipment and located in rich farming communities using the most up-to-date machinery, is the high school at Waterford borough and township, in Erie County, Pennsylvania. The township contains much good farming land, supports prosperous dairying interests, and has the advantage of excellent markets near at hand. The community is an old, established one, and the farming population conservative in its methods. The stone "academy" built in 1822 still forms the principal part of the high-school building. The school board is made up of far-seeing, practical business men, some of whom have had the advantages of foreign travel and extensive business operations. They have inaugurated a workable course in agriculture running through most of the four years, and hope to introduce home economics as soon as public opinion is educated up to it. That the work is exerting a wholesome influence is shown by the fact that the students had tested milk for 40 patrons of the cheese factory up to May 10, 25 of them being farmers with dairy herds ranging from 20 to 40 cows each. In the course of time boys in many of these homes will be able to use Babcock testers owned by their fathers. At present although the farmers have not gained the confidence to test their own seed corn, many have sent in samples to be tested by the class in agronomy. Some sent in shelled corn while others sent lots varying from two ears to three bushels. Four such lots arrived on the day of my visit, May 10. Most of the farmers, however, have not yet been converted into believers in seed testing.

The district comprises about 25 square miles and has a population of 1,500 or more, half of this being in the borough.

An average of 20 students have graduated for several years, 75 per cent of whom later teach in the rural schools. As instruction in agriculture was instituted in the fall of 1904 it will readily be seen that its influence is felt in no small degree in the rural schools of the township. A synopsis of the literary, scientific, and agricultural courses of the high-school curriculum is here given:

CURRICULUM OF THE WATERFORD HIGH SCHOOL

| Literary | Scientific | Agricultural |
|---|---|---|
| | *First Year* | |
| English<br>Algebra<br>Latin<br>Physical geography<br>Arithmetic (two terms) | English<br>Algebra<br>English history (one term)<br>Physical geography<br>Arithmetic (two terms) | English<br>Algebra<br>Plant life (*a*)<br>Physical geography<br>Arithmetic (one term) |
| | *Second Year* | |
| English<br>Algebra (two terms)<br>Latin<br>Civics (one term) | English<br>Algebra (two terms)<br>Zoology and physiology (elective)<br>Bookkeeping, commercial law, and commercial geography (elective) | English<br>Algebra (two terms)<br>Civics (one term)<br>Zoology and physiology<br>Field, orchard, and garden crops (*b*) |
| | *Third Year* | |
| English<br>Physics<br>Latin<br>German | English<br>Physics<br>Chemistry<br>U. S. history (one term)<br>Astronomy (one term)<br>Gray's botany (one term) | English<br>Physics<br>Chemistry<br>Domestic animals (*c*)<br>Soil physics (one term)<br>Gray's botany (one term) |
| | *Fourth Year* | |
| English<br>Geometry<br>General history<br>Latin<br>German | English<br>Geometry<br>General history<br>Chemistry (two terms)<br>Psychology (one term)<br>Geology (one term)<br>Common branches review of (one term) | English<br>Geometry<br>General history<br>Chemistry of soils, plant and animal life<br>Geology (one term) |

*a* " Plant Life " dealt almost entirely with plant physiology, reproduction, and improvement.
*b* Equivalent to agronomy with a small amount of horticulture.
*c* Equivalent to animal husbandry.

The enrollment for the year 1907-8 was 89, distributed as follows:

| | | | | | | |
|---|---|---|---|---|---|---|
| Fourth year, | boys | 6, | girls | 14, | total | 20 |
| Third year, | boys | 2, | girls | 3, | total | 5 |
| Second year, | boys | 4, | girls | 14, | total | 18 |
| First year, | boys | 23, | girls | 17, | total | 40 |
| Unclassified | boys | 5, | girls | 1, | total | 6 |
| Total | boys | 40, | girls | 49, | total | 89 |
| 1905–1906 | boys | 40, | girls | 44, | total | 84 |
| 1903–1904 | boys | 41, | girls | 63, | total | 104 |

One-half of the students are from the township and most of the borough children are of parents who live in town but manage farms they own. About 90 per cent of all the students come under one of these heads, the remaining 10 per cent being children of merchants, laborers, etc.

Only 1 per cent goes to college, though a few have gone to the agricultural college since instruction in agriculture has been introduced into the school.

Seventy-nine in all were registered in the courses in agriculture; 12 in the first-year "plant-life" class, 15 in the second-year "field, orchard, and garden crops" class, including a few fourth-year students who elected it, and 16 in fourth-year agricultural chemistry, which had been preceded by the regular third-year class in chemistry, largely demonstrational in its nature. No class in "domestic animals" was in operation on account of the small size of the third-year class.

The enrollment in these classes taken with the school enrollment given above shows the hold that the work has on the students of all years in the high school.

The largest single class was the one studying poultry. It was a voluntary study not included in the regular curriculum, and carried on after school because a place could not be found for the subject in the daily program. While this arrangement greatly interfered with systematic work, the interest was well sustained. There were over thirty students in the class, the organization of which was inspired by the addresses on poultry delivered before the farmers' institute February 21, 1908. Because of the presence of so many children, some from grades as low as the sixth, the lecturer, Mrs. Allen, stated that she would talk about her hobby,—chickens. She was requested to continue the topic at the evening session. A day or so later in a class studying Shakespeare, taught as it happened by the instructor in agriculture, one of the girls suddenly exclaimed, "Mr. Button, why can't we study poultry?" The idea was so popular that a class was immediately organized. Incubators were lent by patrons, kerosene was furnished by the board, and eggs by the students. One incubator was furnished at half-price by the manufacturer, four workable machines were loaned by farmers, and one that did not operate successfully. This was made the subject of an investigation on faulty construction. There is no manual training in the school, but many of the boys constructed brooders after the specifications given. As this course is unique among all those that came under personal observation, a detailed description of it may not be amiss.

The students worked in pairs, one from the borough with one from the country. Each pair attended to an incubator or to a tray for a week at a time, keeping a daily record of the temperature at morning, noon, and night, of the time of turning the eggs both morning and evening, and of the time the air was allowed to cool. The record also showed the date of testing the eggs and the number "tested out" each time. This was done about every six or seven days. Then after the hatch, the number of chicks and of unhatched eggs was entered on the record. Cardboard sheets about 12 x 18 inches were used for the record. The upper two-thirds was appropriately ruled for the data, while on the lower third were entered the number of hatched and unhatched eggs, the number of the incubator and tray or section, and the signatures of the observers for each week. While each student wrote his name or initials on the eggs he furnished, no attempt was made to have him care for his own eggs.

The English of each of the four years was in charge of the instructor of agriculture, so that a closer correlation of the two lines of work was possible than I found in any other school visited. Not a little of the work in English was based on the activities of the agricultural classes. One of the efforts in the senior rhetoricals, which occupied the time of the opening exercises was based upon an agricultural theme. Some of the composition work was directly related to the readings and lectures of the agricultural classes, so that their notes took the form of a formal exercise in English. I inspected a large number of essays on the same topic, poultry houses, handed in a day or so before my coming. The guide for the course was Lockwood's "Lessons in English." Each essay was prefaced, of course, with an outline. From this lot one essay was selected because the outline was really a synopsis of the essay, such as would present in brief compass a good notion of the essay itself and of the topic as presented in the agricultural class. The author was the daughter of a hotel keeper, and without previous experience with poultry.

This synoptical outline is here given as written:

## POULTRY HOUSES

*Houses and Inclosures*

1. The systems of housing are:
   (a) Colony plan, and
   (b) The continuous apartment house.
2. The location and elevation of houses should be:
   (a) Facing the south, and
   (b) On elevation with natural drainage away from buildings.
3. The proper time to build is:
   (a) During spring or early summer so as building will have time to dry out.
4. The size and dimensions of a house:
   (a) Depends on the number of fowls to be kept, i.e., about five square feet to every chicken.
   (b) Should be square, and
   (c) As low as possible, as a low house is more easily warmed than a high one.
5. Foundation walls:
   (a) When permanent should be made of brick, stone or concrete or
   (b) When large stones or brick are not available use small stones.
6. The different types of roofs are:
   (a) The single pitch,
   (b) The gable roof, or double pitch, and
   (c) The combination.

7. Floors may be made of:
    (a) Earth,
    (b) Wood, or
    (c) Cement.
8. Windows.
    (a) Should be placed high and vertical in the walls, and should be
    (b) About eight by ten inches in size of pane.
9. The quality of construction:
    (a) Should be serviceable, fairly roomy, well lighted, and well ventilated, and
    (b) Constructed according to the location and climatic conditions.

*Interior Arrangements*

1. Roosts:
    (a) Seven to 10 inches should be allowed each fowl,
    (b) A perch 16 feet long will accommodate 20 chickens.
2. Nests:
    (a) Should be placed in a dark place.
    (b) Partitions should be between nests.
    (Partitions) low enough to permit hens to go from one nest to another, and
    High enough to prevent eggs from rolling from one nest to another.
3. Dust boxes:
    (a) Should consist of fine, light, dry dust,
    (b) Sandy loam, or
    (c) Road dust mixed with coal or wood ashes.
4. The drinking fountain:
    (a) Should be large enough to hold water for twenty-four hours, and
    (b) Elevated from the floor.
5. Feed trough and grit boxes:
    (a) Should be constructed so that the chickens cannot get in them with their feet.
6. Ventilate with cloth curtains over openings.

While no class in animal husbandry was organized in 1907-8 for the reason already stated, the use of the Babcock milk test was taught to all classes. The tester purchased by the school was an eight-bottle, cased machine, costing $10 or $11. The center is interchangeable so that a twelve bottle-center may be substituted. The dairymen send samples for testing to "keep tab" on what they take to the creamery. As the creamery paid by the test and the cheese factory did not, the latter got the thin milk and was compelled also to adopt the Babcock test standard. Perhaps the fact that the proprietor of the cheese factory was a member of the board accounts for the board's appreciation of a tester as a piece of school apparatus. Some of the farmers were having their herds tested through a series of weeks for the purpose of thinning out their poor stock. The pupils made duplicate tests of those made by the instructor, kept the records, and got much practice in that way. The school also had the use of a cream separator valued at $85.

Corn judging was a regular part of the work of the second year class, although some of the testing was done by the first year students. The ears of a lot brought in were spread out on a cupboard shelf and numbered. Six kernels were taken from each ear, two each from the tip, middle, and butt. The seed tester was made of a soap box with a layer of sawdust covered with muslin ruled off into squares. The box was deep enough to permit of several layers of seeds being placed in it, but since no provision had been made for rigid trays, this method was found to have no advantage if the seeds were to be inspected before sprouting.

From forty to fifty farm journals are received and filed each week. Many of the seniors, in connection with their work in English, write to some of these journals and to the daily papers of Erie.

Much of the equipment has been purchased with funds raised by the students. Six years ago, the school raised by entertainments $75, which was duplicated by the board. Eighty-five dollars was raised in a similar manner the year of my visit.

The per capita cost of the school is $27. The tax rate of 4 mills for the borough for the common schools, has risen to 8 mills since it took over the academy. The total borough rate is 16 mills on an assessed valuation estimated at 66 per cent or over of the actual valuation.

The school is unfortunate in having different instructors in agriculture every year, as this prevents a plan being carried out consistently from year to year. As it is, the instructors in agriculture are paid a half more than other assistants, but after gaining a year's experience in teaching they are able to command better salaries than the board feels able to pay. The work at the time of this visit was in charge of Mr. H. F. Button, a graduate of the New York State College of Agriculture at Cornell University, now director of the agricultural and normal training high school established at Manassas, Va. His immediate predecessor, Mr. H. O. Sampson,[5] a graduate of the Iowa State College of Agriculture, left to enter the service of the Department of Agriculture, while another predecessor joined the staff of an agricultural college.

## High Schools Teaching Agriculture Only Incidentally

### *Odell (Ill.) High School*

In addition to the schools already described, a few others were visited. Three of these were academies in New Hampshire and Vermont with long and honorable histories as classical preparatory schools, which had more recently been giving serious thought to the question of adapting themselves to modern industrial conditions. Their efforts had not produced results at that time which can be dealt with to advantage here.

Of several high schools claiming to give instruction in agriculture, not in special courses, but "incidentally" in the usual science courses, the one at Odell, Illinois, merits some notice, as it was the finest example seen of this type of instruction. This type of work has called forth much derision from the advocates of agricultural education, and too often deservedly so. But the students at Odell seemed to derive more benefit from their work than do the students in the majority of schools numerously reported from different states as requiring all students to take the course in agriculture.

Odell is a village of 1,200 in the rich corn-producing central part of the state. The school is not large, enrolling only 27 students. The courses in botany and zoology were each eighteen weeks in length. In the former course were considered soils and their cultivation, and grain farming versus live stock farming; and in the latter, the topics of birds and injurious insects.

[5] See Yearbook of the Department of Agriculture for 1905, pp. 264–4, for an illustrated description of the work of Mr. Sampson, especially in animal husbandry, at the Waterford High School.

In physics the pupils studied specific heat as applied to soils, the color of soils, etc. Some of the students' compositions on corn epitomized their work in the field, laboratory, and reference work in connection with botany. The larger part of the essays did not deal with the structure of the stem and flower, as might be supposed, but with questions concerning the preparation and cultivation of the soil, treatment of the growing and matured crop, marketing, and corn products. The class made use of a commercial corn tester, valued at $11.50. It uses earth in trays covered by a series of metal cross strips dividing the area into one and one-half inch squares, 210 in number. The two trays give a capacity of 420 ears. A wire rack goes with it, in which can be placed the ears from which the kernels are taken. The rack has figures across the top and letters down the side corresponding to a similar notation on the tester. The tester was lent by Mr. Vincent, the president of the board, a young business man who is a graduate of the Iowa State College of Agriculture and is much in sympathy with agricultural instruction.

Mr. Vincent has also presented eight acres for experimental purposes. The school board is financing this work for the benefit of the community and not for the immediate use of the school, though the pupils will observe the work. The work is under the direction of Dr. C. G. Hopkins, of the University of Illinois, as is the experiment field of the John Swaney School. Superintendent Vaughn has immediate charge of the work. The board expected to devote three acres to raising fine seed corn from $5 stock. They hoped to sell this at a price sufficient to defray the expense of conducting the plats. Oats and clover were in one plat, and a new strain of Montana oats was being tried in another. Four series, each with eight one-tenth-acre plats, were being started with the different crops of a four-year rotation—corn, oats, wheat, and clover. Four plats in each series were being handled according to the grain farming system, and four according to the live-stock farming system. The four plats of each sub-series will receive different treatment. Accurate records will be kept by the superintendent, and the pupils of the grades will get material from the experiment farm for many "real" problems in arithmetic, calculating yields, percentages, profits, and the like.

# CHAPTER IV

## ADMINISTRATION, EQUIPMENT, AND METHODS

The immediate problems confronting authorities who think of introducing agriculture into their schools are those of expense, of the relation of the new study to the present curriculum, and the problem of the teacher. Some of the more important administrative questions concern the time given to agriculture, its place in the course of study, its relation to the other sciences and to the curriculum as a whole, and the teaching force that is to handle the subject.

The data given here seem extensive enough to indicate current practice. In some cases the present usage represents the result of several years of experience; in most cases it represents tentative efforts of schools but recently introducing agriculture, or of schools themselves organized but two or three years. It would be difficult in most cases to devise a way to determine the relative merit of the different procedures.

When data from the Alabama district schools are included in tables the fact will be specially mentioned.

### Time Given to Instruction

The series of tables under this heading shows in several ways the amount of time given the subject. Owing to the well-known unreliability of the "average" as giving an accurate notion of conditions, a broad distribution is made of the schools in a way to show the general tendencies at a glance. The distribution by weeks shows how *extensive* the subject is in the respective schools, while the distribution by minutes per week shows how *intensive* it is. The two of course do not necessarily go together.

TABLE 18

DISTRIBUTION OF SCHOOLS ACCORDING TO THE NUMBER OF WEEKS GIVEN TO AGRICULTURE

| Weeks per year | No. of schools | Weeks per year | No. of schools |
|---|---|---|---|
| 6 | 1 | 24 | 1 |
| 8 | *a*1 | 26 | 0 |
| 10 | 2 | 28 | 0 |
| 12 | 7 | 30 | *c*2 |
| 14 | 1 | 32 | 17 |
| 16 | *b*25 | 34 | 3 |
| 18 | 82 | 36 | 12 |
| 20 | 8 | 38 | 5 |
| 22 | 0 | 40 | 4 |
| | | | 171 reporting—all courses one year or less. |

*a* Includes one of 9 weeks.
*b* One of 17 weeks.
*c* One of 31 weeks.

There were probably 50 other schools, mostly in Ohio, Missouri, and Nebraska, teaching agriculture 36 weeks or less at the time the above reports were received.

In addition to the above, the Hamilton County High School at Tyner, Tenn., and the State Preparatory School at Tonkawa, Okla., have a two-year course in operation. The Guthrie County High School at Panora, Iowa, the joint township high school at Waterford, Pa., and the high school at Nephi, Utah, have three-year courses, while two or three New Hampshire academies have instituted three-year courses to follow an introductory course in biology. Four-year courses are in operation at the Magnolia Township High School near McNabb, Ill., the Cecil County High School at Calvert, Md., the Township High School at Petersham, Mass., the Model High School at North Adams, Mich., the Beaverhead County High School at Dillon, Mont., and the eleven congressional district high schools of Alabama. In most cases strictly agricultural studies do not extend through every one of the years mentioned, as usually in one or more of the years it is but a half-year study.

TABLE 19

DISTRIBUTION OF SCHOOLS ACCORDING TO THE NUMBER OF MINUTES PER WEEK GIVEN TO AGRICULTURE

| Minutes per week | No. of schools | Minutes per week | No. of schools |
|---|---|---|---|
| 1– 20 | *a*1 | 301–320 | *a*1 |
| 21– 40 | *a*3 | 321–340 | *a*1 |
| 41– 60 | 0 | 341–360 | *a*1 |
| 61– 80 | 0 | 361–380 | 0 |
| 81–100 | *a*2 | 381–400 | *a*5 |
| 101–120 | *a*4 | 401–420 | 0 |
| 121–140 | *b*4 | 421–440 | 0 |
| 141–160 | *c*19 | 441–460 | 1 |
| 161–180 | 10 | 461–480 | 0 |
| 181–200 | *a*58 | 481–500 | 0 |
| 201–220 | 5 | 501–520 | 0 |
| 221–240 | *d*15 | 521–540 | 0 |
| 241–260 | 2 | 541–560 | 0 |
| 261–280 | *a*5 | 561–580 | 0 |
| 281–300 | 3 | 581–600 | *a*1 |
| | | | 141 schools —including those of more than one year's work if reporting. |

*a* Usually denotes highest number of minutes.
*b* Usually denotes 135 minutes.
*c* Usually denotes 150 minutes.
*d* Usually denotes 225 minutes.

A closer distribution gives the following:

| Minutes per week | No. of schools | Minutes per week | No. of schools |
|---|---|---|---|
| 1–100 | 6 | 301–400 | 6 |
| 101–200 | 88 | 401–500 | 1 |
| 201–300 | 29 | 501–600 | 1 |
| | | Total | 141 |

The usual number of recitations per week is five, and the usual length of the recitation is 40 minutes, which gives 200 minutes as by far the most frequent case; 30-minute recitations in schools, with small teaching force gives 150 minutes per week; five double periods of 80 minutes each gives 400 minutes per week, while the smaller number of double periods gives the amounts occurring less frequently.

It is evident that a large number of minutes per week for a large number of weeks results in a heavy course and vice versa.

The following table is the result of multiplying the minutes per week by the number of weeks, and gives a truer measure than either of the other tables. No effort was made to learn the times per week the subject was given, as the point would be of but little value, as would also be the length of the recitation.

TABLE 20

DISTRIBUTION OF SCHOOLS ACCORDING TO THE TOTAL NUMBER OF MINUTES PER ANNUM GIVEN TO AGRICULTURE

| Minutes per annum | No. of schools | Minutes per annum | No. of schools |
|---|---|---|---|
| 1- 1000...... | 3 | 9001–10000..... | 1 |
| 1001- 2000...... | 10 | 10001–11000..... | 1 |
| 2001- 3000...... | 18 | 11001–12000..... | 0 |
| 3001- 4000...... | 52 | 12001–13000..... | 0 |
| 4001- 5000...... | 13 | 13001–14000..... | 0 |
| 5001- 6000...... | 9 | 14001–15000..... | 2 |
| 6001- 7000...... | 12 | 15001–16000..... | 0 |
| 7001- 8000...... | 9 | 16001–17000..... | 1 |
| 8001- 9000...... | 4 | | |
| | | | 135 |

All of these 135 courses run one year or less.

The most frequent number of minutes in the above schools are:

2,400 minutes in................................8 schools.
3,200 minutes in................................7 schools.
3,600 minutes in................................37 schools.
4,000–4,050 minutes in..........................8 schools.
6,400 minutes in................................9 schools.
Including over half of the 135.

Cases of 2,400 minutes usually mean 12 weeks of 40 minutes daily recitations.

Cases of 3,200 minutes, 16 weeks of 40 minute recitations.
Cases of 3,600 minutes, 18 weeks of 40 minute recitations.
Cases of 4,000 minutes, 20 weeks of 40 minute recitations.
Cases of 4,050 minutes, 18 weeks of 45 minute recitations.
Cases of 6,400 minutes, 32 weeks of 40 minute recitations.

## RELATION OF AGRICULTURE TO THE CURRICULUM AND TO THE OTHER SCIENCES IN THE CURRICULUM

No less troublesome a question than that of time, is the place of the study when continued through a single term or year.

Some educators advocate its coming after the other sciences, which would place it in the fourth year, or at the lowest, in the third. Others contend that it should function as the course in introductory science for the entering high-school students.

TABLE 21

DISTRIBUTION OF SCHOOLS ACCORDING TO THE YEAR IN WHICH AGRICULTURE IS GIVEN

| | No. of schools |
|---|---|
| First year | *a*49 |
| Second year | 31 |
| Third year | 14 |
| Fourth year | 10 |
| Total | 104 |
| First or second year | 17 |
| First, second, or third | *b, c*2 |
| First, second, third, or fourth | 7 |
| Second or third | 5 |
| Second, third, or fourth | 2 |
| Third or fourth | 7 |
| First or fourth | 1 |
| Total of 145 courses of one year or less | 41 |
| First and second years | 2 |
| First, second, and third | 1 |
| First, second, third, and fourth | *d*6 |
| Second, third, and fourth | 2 |
| Third and fourth | 1 |
| Schools reporting | 12 |
| | *e*157 |

*a* Lake Charles, La. Eighth grade is the first year of the high school.
*b* Athens, Ga. Eighth, ninth, and tenth grades are the first, second, and third years of high school.
*c* Holland, Ind., admits pupils of the eighth (grammar) grade as well as first and second year high-school students.
*d* Fifteen schools including the Alabama schools.
*e* One hundred sixty-nine schools including the Alabama schools.

The advocates of physical geography some years ago, claimed this as one of the strong points of their favorite. However, it has been disappointing in that the number and variety of scientific principles involved are too restricted to introduce to much of anything but geology, a study that has almost vanished from high schools. Consequently, physical geography must stand on the intrinsic worth of the information of the subject. Table 21 shows in what year agriculture is given by schools offering single courses. It will be seen that one hundred and four schools give it a definite place in a single year. Forty-one allow it to be taken by pupils in two or more high-school grades, and twelve schools, in addition to the eleven congressional district high schools of Alabama, give different agricultural courses in more than one year.[1]

The above table shows that in 76, or over one-half, of the high schools teaching agriculture for one year or less, the pupils may or must take it in the first year, while in 64 or almost half of these schools, they may or must take it in the second year. In only 69 schools are first-year students excluded, and in only 31 schools are first and second year students excluded. Twenty-nine schools open the subject to two succeeding classes, as first and second, second and third, third and fourth. In most cases this is due to the doubling up of classes to economize teaching force, a device not confined to agriculture.

One of the chief criticisms to be made on the administration of small high schools is, that this principle of doubling classes is not used oftener. It would prevent the teacher's energy from being dissipated over so many small classes, and would give the classes the added inspiration that comes from numbers. The writer has seen teachers in rural high school spending forty minutes with the one pupil composing a fourth year physics class—most expensive instruction when one considers that the entire third year class of three or four might have been taught at the same time with no more trouble and with much greater effectiveness.

Eighty-one schools report the subject as being required; in one it is required of the boys. Sixty-three schools offer it as

[1] At the end of the school year 1908–9, the United States Department of Agriculture had reports from 47 high schools with four-year courses, 11 with three-year courses, 38 with two-year courses, 90 with one full year's work, and 123 with part-year courses. (See Annual Report of the Office of Experiment Stations, 1909, pp. 307–8). Most of these schools are included in Table 1, first and second columns, and in Table 2, pages 24 and 25.

an elective. It must be remembered that the term "elective" covers a wide range of freedom. In some schools it is an "alternative" rather than a free elective, to be taken in place of, say, Latin. It is required in the "agriculture course" as opposed to the classical course in two or three schools. One school requires it in the science course. It is more apt to be required in small schools, especially in those with a two or three-year course, than in large schools. The reason is obviously one of economical use of the teaching force. When two classes, as the first and second, are consolidated or "doubled," it is evident that a greater opportunity can be given in the way of electives in this or in other branches, as that much more of the instructor's time is released for other recitations.

The content of the subject, scarcely less than the method of teaching it, is modified by the kind and amount of science preceding it or being studied parallel to it. Table 22 shows this relationship between agriculture and the other sciences. The

TABLE 22

DISTRIBUTION OF SCHOOLS ACCORDING TO THE SCIENCES PRECEDING AGRICULTURE

| Agriculture preceded by | No. of schools |
|---|---|
| Biology | *a*1 |
| Botany | 33 |
| Chemistry | 6 |
| Geology | 3 |
| Physics | 16 |
| Physical geography | 72 |
| Physiology | *b*17 |
| Zoology | 5 |
| 153 cases reported | 108 |
| Reporting parallel science | *c*3 |
| Reporting no parallel science | 29 |
| Not reporting | 9 |
| Total | 149 |

*a* In many cases the botany partly precedes and partly follows the agriculture, which is then taught in the winter months. This is probably true in a much larger number of cases than I have record of.

*b* There seems to be good reason to suspect that in many cases this was not high school physiology.

*c* One case of botany preceding or paralleling agriculture was counted as preceding it.

large number of cases is due to the fact that some schools report more than one science as preceding agriculture.

One school reporting botany and zoology, or physics was counted three times. One school reporting agriculture as a fourth year study to be preceded by "any laboratory science," could not be counted in the above table.

TABLE 23

DISTRIBUTION OF SCHOOLS ACCORDING TO THE AMOUNT OF SCIENCE PRECEDING AGRICULTURE

Counted another way the reports show:

| | |
|---|---|
| 29 schools reported | 0 preliminary science. |
| 70 schools reported | 1 preliminary science. |
| 26 schools reported | 2 preliminary sciences. |
| 11 schools reported | 3 preliminary sciences. |

With the possibility of complications due to parallel studies being reported as preliminary.

TABLE 24

DISTRIBUTION OF SCHOOLS WITH FIRST-YEAR AGRICULTURE PRECEDED BY ANOTHER SCIENCE

| Of those schools having agriculture in the first year | Schools |
|---|---|
| Biology is reported as preliminary in | 1 |
| Botany is reported as preliminary in | 2 |
| Geology is reported as preliminary in | 1 |
| Physical geography is reported as preliminary in | 24 |
| Physiology is reported as preliminary in | 7 |
| No science reported in | 26 |
| Not reporting | *a*12 |
| Total number schools reporting | 73 |

*a* Probably most of these schools may be counted as having no science preliminary to agriculture.

This table indicates (1) how largely botany is a second year study, or (2) how often it follows agriculture in the first year, as it sometimes does, when it is often little more than plant analysis.

## THE TEACHING FORCE

Since the subject of agriculture is new and few teachers are qualified to teach it, we find it handled often by the superintendent or principal, i. e., the principal of the entire village or township system, though occasionally by one who has jurisdiction of only the high school. The superintendent sometimes takes the work because he is the only man in the system, the only one with any practical farm or laboratory experience, or because of the unsympathetic attitude of his science teacher. Quite often he is the science teacher.

Table 26 shows that but two of these teachers, whether superintendents or not, teach no other branches, and that only 31 teach three other branches or less. The qualification of these persons to teach agriculture is considered in the chapter on Salaries and Preparation.

TABLE 25

OFFICIAL POSITION OF THE INSTRUCTOR IN AGRICULTURE, (INCLUDING THE SEVEN ALABAMA SCHOOLS REPORTING)

| | |
|---|---|
| Taught by principal or superintendent | *a*104 |
| Taught by high-school assistant | *b*45 |
| Taught by special agriculturist | 11 |

*a* Three of the principals were trained in college to be agriculturists.
*b* Two of the high-school assistants share the work of agricultural instruction with their superintendent.

It need not be said that a thoroughly trained teacher can teach a large number of classes more efficiently than one not trained. The teacher well trained in the sciences might also be expected to take up agriculture more easily than one not so trained. In order to be well taught, agriculture requires time for setting up experiments, as certainly as do other sciences.

TABLE 26

DISTRIBUTION OF TEACHERS ACCORDING TO THE NUMBER OF CLASSES TAUGHT IN ADDITION TO AGRICULTURE

| Number of classes besides agriculture | Number of teachers | Number of classes besides agriculture | Number of teachers |
|---|---|---|---|
| 0 | 2 | 11 | 5 |
| 1 | 1 | 12 | 3 |
| 2 | 11 | 13 | 0 |
| 3 | 17 | 14 | 0 |
| 4 | 22 | 15 | 2 |
| 5 | 30 | 10 | 0 |
| 6 | 15 | 17 | 0 |
| 7 | 15 | 18 | 1 |
| 8 | 10 | "Many" | 2 |
| 9 | 3 | | |
| 10 | 5 | Total | 144 |

It also needs time, as the physical sciences do not, for trips to the field, stock farms, factories for canning and for milk products, etc. The instructor who must teach many other classes or who must spend long hours in the school room can not be expected to teach agriculture as it should be taught.

Tables 26 and 27 show how the time of high-school teachers of agriculture is spent.

TABLE 27

DISTRIBUTION OF TEACHERS ACCORDING TO THE NUMBER OF HOURS ON DUTY IN SCHOOL

| Hours on duty | Number of teachers | Hours on duty | Number of teachers |
|---|---|---|---|
| 2 | 2 | 6 | 45 |
| 2½ | 0 | 6½ | 3 |
| 3 | 8 | 7 | 6 |
| 3½ | 0 | 7½ | 1 |
| 4 | 14 | 8 | 4 |
| 4½ | 3 | All | 2 |
| 5 | 19 | | |
| 5½ | 12 | Total | 119 |

As might be expected, only 6 teachers of the 31 teaching three classes or less besides agriculture are not superintendents or principals. But the 6 are not, as one might expect, the special agriculturists.

The inevitable influence of the numerous classes is shown by the time given per week. For the reports quite generally show that agriculture receives the least amount of time in those schools reporting the larger number of recitations to the teacher.

TABLE 28

TOTAL NUMBER OF CLASSES ASSIGNED TO TEACHER OF AGRICULTURE, AND TIME SPENT DAILY IN AGRICULTURE

(Five items of Table 26.)

| Number of classes | Number of teachers | Time spent—minutes |
|---|---|---|
| 10 | 5 | 120–150 |
| 11 | 5 | 100–180 |
| 12 | 3 | 90–150 |
| 15 | 2 | 100–125 |
| " Many " | 2 | 135–180 |

One overworked North Carolina teacher with 18 classes manages to spare 20 minutes per week to agriculture, and succeeds in the course of 20 weeks in demonstrating most of the experiments in the text. Little wonder, however, that the attitude of the class is reported as "indifferent."

Besides the difficulties already cited, a few others of an administrative sort were mentioned, such as, "the shortness of the season," "unsuitable winter season" (given twice), "the necessity of carrying experiments home at night on account of the building getting cold," "immaturity of the pupils," "poor work done in the grammar grades," "size of the class," "too many other studies," "determination of the place of agriculture in the course," "too heaviness of the course."

## Experimentation and Field Work

A point in favor of the modest course in agriculture introduced by many of the smaller high schools, is that it requires little equipment in addition to that used in the sciences usually taught. It requires no more elaborate laboratory facilities. In fact, its initial installation requires less expense than either physics or chemistry, while botany and zoology can get along with no less apparatus, and the use of the miscroscope renders their cost much more. If properly taught, physical geography should have as much in the way of apparatus, maps, and models, although this study has freed itself less than any other science taught in small high schools from the stigma of being a book science.

Much of the apparatus distinctively agricultural can be made by any boy handy with carpenters' tools, and should, in fact, often does, furnish the direct motive for the manual training work. The home or village dump will supply most of the tin cans, pickle bottles, and jars. It is significant that 6 schools reporting "no difficulties" also reported "no money spent for agricultural apparatus." Their reports indicate that the work was not of the book type. Of 115 reporting their chief difficulties, apparatus and laboratory facilities were specifically mentioned by 31, while 13 others mentioned "equipment," "facilities," and "money," which might apply equally to difficulties standing in the way of gardening. Seven of the 31 had spent respectively $10, $12, $20, $25, $30, $40, and $250. A school board that will introduce agriculture and refuse a moderate amount of apparatus, say $50 worth, will treat the other sciences in the same way.

Fifty-six schools reported expenditures, loans, and gifts of $4,833. This included a gift of an $800 greenhouse at Petersham, Mass., and $1,000 spent at Dillon, Mont., partly in equipping a blacksmith shop. This leaves $3,033 reported by the remaining 54 schools and by Petersham for some apparatus. Eight schools reported $887 for outdoor work, leaving $2,146 reported by 50 schools for indoor work, 4 schools reporting on both items.

TABLE 29

PURPOSES TO WHICH THE AVAILABLE FUNDS ARE DEVOTED

The $2,146 was reported under the following headings:

| | |
|---|---|
| By 4 schools, value of loans of sample seeds, apparatus, etc.... | $207 |
| By 5 schools, value of gifts, of seeds, apparatus, etc........... | 460 |
| By 16 schools, money spent for unspecified purposes, probably for the same purposes as already mentioned............... | 497 |
| By 33 schools, money spent for sample seeds, apparatus, etc... | 982 |
| | $2,146 |

Of the above amount $120 was spent for four compound miscroscopes and a balance. Three other items included a Harvard balance and thermometers. The microscopes are biological apparatus rather than agricultural, while the other articles are as much physical as agricultural apparatus. But when biology and physics are not present in the course of study of the small high school, there is no limit to the degree to which their content may be incorporated into the agricultural course and their apparatus appropriated to its purposes.

54 schools reported no money spent for apparatus other than that used in the other sciences.

2 reported only home-made apparatus.

1 reported a good equipment for general science.

1 mentioned that a fruit experiment station was near the school.

1 reported $50 spent for apparatus for all the sciences.

1 reported that it could borrow agricultural apparatus.

1 has the use of all kinds of farm machinery handled by the local dealers in farm implements.

TABLE 30

DISTRIBUTION OF SCHOOLS ACCORDING TO AMOUNTS OF LOANS, GIFTS, AND EXPENDITURES FOR ALL AGRICULTURAL PURPOSES

| | No. of Schools |
|---|---|
| $25 or less | 28 |
| 26 to 50 | 16 |
| 51 to 100 | 3 |
| 101 to 200 | 3 |
| 201 to 300 | 3 |
| 301 to 400 | 1 |
| 401 or more | 2 |
| 10 to 50 (estimated) | 10 |

The estimates for the last item were based on general description of the material reported, although no definite value was placed on it.

It may be of interest to know what apparatus the schools have mentioned in their reports.

10 schools reported soil tubes, pans, or boxes.
12 schools reported samples of grain, seeds, soils, and fertilizers.
24 schools reported milk testers.
2 schools reported milk separators.
2 schools reported incubators.
2 schools reported agricultural implements.
1 school reported a spraying machine.
1 school reported grafting knives.

Doubtless many more schools could have reported one or more of these items had they cared to take the time.

Nearly all the schools used texts containing a number of experiments that are practicable for indoor use. Several of the schools based their experimentation work on some manual, as Office of Experiment Stations Bulletins Nos. 186 and 195, and Circular No. 77; Bulletin No. 2 of the Minnesota Agricultural College; Bulletins No. 1 and 2 on Agricultural Education, Miami University; and State Manual of Nebraska. Other schools report these as basis of extra experiments performed.

As a measure of the efficiency of instruction in the general high school, the amount and character of the experimentation work done is probably as good as any, just as the practical farm plat demonstration may serve as an index of the work done in the technical agricultural high schools. The accompanying table gives a notion of this indoor work.

TABLE 31

AMOUNT OF EXPERIMENTATION WORK

106 schools report doing practically all the experiments in the text or manual used.
47 schools report doing experiments outside of text.
26 schools report doing "a few," "some," or "not many" experiments.
46 schools report doing unspecified laboratory work. (*a*)
14 schools report doing no laboratory work.
10 schools omitted this item on blanks filled out. (*b*)

*a* These were reported on preliminary postal card inquiries, no fuller data being later supplied.
*b* To these 10 should be added as many more schools whose preliminary report indicates such poor work that no further inquiry was made.

Only four schools reported the amount expended on gardens. \$40 for land, \$5 and \$250 for tools, and \$260 for garden, total \$555.

TABLE 32

WHO PERFORMED THE EXPERIMENTS

18 schools report the work done chiefly by the pupils individually.
18 schools report the work done chiefly by the pupils in groups.
32 schools report the work chiefly demonstrated by the teacher.
15 report the work done equally by pupils individually and in groups.
14 schools report the work done equally by the pupils individually and the teacher.
23 schools report the work done equally by the pupils in groups and by the teacher.
12 schools report the work done by all three.

132 reporting on this point. The others did not report or did no laboratory work.

Three schools report lack of land as their chief difficulty, 2 report lack of gardens, 2 lack facilities for outdoor practice, and 1 the inability to get land on a long lease, a total of 8 reporting their chief difficulties along these similar lines.

TABLE 33

SCHOOLS REPORTING GARDEN WORK

28 schools report school gardens without qualifications.
9 schools report school gardens for training class (Nebraska).
6 schools report "a little" or "irregular" gardening done, mostly with flowers.
160 schools report "none" or fail to report.
35 schools report "home gardens" without qualifications.
1 school reports "home garden" for training class (Nebraska).
74 schools report "none."

Over one-third of the schools reporting some kind of home garden work are in Ohio, where the State University in 1906-7 and 1908 made a well organized effort to stimulate the movement among the few high schools then teaching agriculture, by sending out seed, printed directions, and blanks for reports. The effort was discontinued as soon as the work seemed fairly well established, and the schools have since been encouraged to continue in the work without outside aid and to carry on breeding experiments, etc., with home grown seed. In some cases the high schools seem to have continued and enlarged this work, in other cases it has been continued by grade children, while still other schools have allowed this line of agricultural work to lapse. Similar work is carried on extensively by school children of the grades, or independently by the local school itself, under the patronage of the county superintendents, the county fairs, granges, agricultural colleges, and state departments of education in New York, Ohio, Indiana, Illinois, North Dakota, Nebraska, Kansas, Oklahoma, Tennessee, and Georgia.

In the North the result is the "corn contests," "flower shows," and "home economics exhibits"; in the South it is the "corn, cotton and chicken contests."

There is not the evidence, however, that the friends of education would like to see, to show that these movements are a part of the actual daily life of the school room, furnishing the problems for solution therein, and material for the work in English, arithmetic, and geography.

A surprisingly small number report taking trips to stock farms, creameries, cheese, butter, or canning factories—only fifteen in all, and two-thirds of these in Nebraska. Three teachers cite "lack of time for field trips" as their chief difficulty, while one laments that no "fields are located conveniently for observation."

## Text and Reference Books

The less of special preparation a teacher has, the more truly the character of the work done is likely to be represented by the text used. By keeping in mind the preparation of the teachers indicated in the next chapter, and the number of different classes handled by the agricultural teachers as shown in the present chapter, one who is conversant with the texts mentioned in Table 34 may gauge somewhat approximately the grade of the agriculture now taught. It should be said, however, that some of the texts of more truly high-school grade, as those by Ferguson and Lewis, or Jackson and Daugherty, both suitable for the lower years, and the more advanced text by Warren, have not been on the market as long as those of purely elementary grade; also that many teachers recognize the unsuitable character of some of the books most widely used as high-school texts. Some of these reported that their pupils regarded

TABLE 34

TEXT BOOKS REPORTED AS USED

| Author and title | References | Texts in high schools | Texts in training classes—Michigan, New York and Wisconsin |
|---|---|---|---|
| Bailey: Principles of Agriculture | 18 | 39 | 5 |
| Bessey, Bruner, and Swezey: New Elementary Agriculture | 4 | 2 | .. |
| Burkett, Stevens & Hill: Agriculture for Beginners | 8 | 71 | 20 |
| Ferguson & Lewis: Elementary Principles of Agriculture | 1 | .. | .. |
| Goff & Mayne: First Principles of Agriculture | 7 | 23 | 19 |
| Goodrich: First Book of Farming | 5 | 2 | .. |
| Hatch & Haselwood: Elementary Agriculture & Practical Arithmetic | 1 | .. | 11 |
| James: Agriculture | 5 | 6 | 2 |
| Jackson & Daugherty: Agriculture through Laboratory and School Garden | 6 | 19 | .. |
| " King " | .. | .. | 2 |
| Sheppard & McDowell: Elements of Agriculture | 1 | 3 | .. |
| Voorhees: First Principles of Agriculture | .. | 1 | .. |
| Total | | 166 | 59 |
| Several texts reported used | | 4 | 4 |
| No text used | | 5 | 5 |
| Report on texts omitted | | 6 | 9 |
| Not reporting fully | | 56 | 12 |

the texts in question as "kindergarten books," or expressed in other terms the idea that the pupils craved something more suited to their capacities.[2]

Two hundred thirty-three high schools or teachers' training classes, most of them in high schools, reported the name of one text each, 10 reported that none was used, 8 reported that "several" were used, 15 omitted this item from reports otherwise rather complete, and 56 schools sending in very incomplete reports omitted this item.

Fifty titles were reported as reference books 141 times in all. Of these, 8 were public school texts, reported 47 times, 5 of which were grammar school texts, reported 24 times.

TABLE 35

BOOKS REPORTED AS REFERENCES FIVE TIMES OR MORE

| Author and title | Times reported |
|---|---|
| Bailey: Principles of Agriculture | 17 |
| Burkett, Stevens & Hill: Agriculture for Beginners | 8 |
| Goff & Mayne: First Principles of Agriculture | 7 |
| Goodrich: First Book of Farming | 5 |
| Hunt: Cereals of America | 7 |
| James: Agriculture | 5 |
| King: The Soil | 8 |
| " (unspecified) | 4 |
| Roberts: The Fertility of the Land | 5 |

Farmers' bulletin 44 was reported 6 times, while bulletins 22 and 28 were reported 5 times each. Office of Experiment Stations bulletin 186 (now superseded by Farmers' bulletins 408 and 409) was reported many times as text, reference, and laboratory manual.

[2]According to a report by G. A. Bricker on high schools in Ohio teaching agriculture in 1910-1911, issued as this work is going to press, the texts used in 243 high schools were as follows: Bailey, 8; Byrkett, Stevens and Hill, 56; Goff and Mayne, 30; Jackson and Daugherty, 26; Warren, 60; Wilkinson, 55; five others, 8. The last two named have been issued since receipt of the replies from which Table 34 was compiled. Warren's text has also been introduced into a large number of schools offering agriculture in the upper years of the curriculum.

# CHAPTER V

## PREPARATION AND SALARIES OF TEACHERS OF AGRICULTURE IN THE HIGH SCHOOLS

### Preparation and Practical Experience of the Teachers

Considerable information is at hand as to the preparation of the persons teaching agriculture in the school studied. In the questionaire sent out the matter was formulated thus: "Special preparation of the high school instructor for teaching agriculture, either theoretical, scientific, or practical? (Indicate nature of this) Is he a college or normal school graduate?"

Some respondents attempted to answer all of the first question by simply writing "yes," "all three," or by underscoring one or more. If the teacher was a college graduate, any "scientific" preparation mentioned has been considered to be college science. Few gained their theoretical knowledge in agricultural colleges, most of them getting it through reading. The analysis of the returns is shown in Table 36.

Table 36

Preparation of 182 High-School Teachers of Agriculture

| | |
|---|---|
| Number claiming to have special preparation or qualifications | 143 |
| Number disclaiming any special preparation or qualifications | 12 |
| Number failing to report special preparation or qualifications | 19 |
| Number reporting on preparation, scholarship, or graduation | 174 |
| Number failing to report on any of these items | 8 |
| Total number of reports used in calculations | 182 |
| Number reporting practical farm experience | 77 |
| Number reporting college courses in agriculture | *a*29 |
| Number reporting college courses in science | *b*41 |
| Number reporting normal courses in agriculture | *c*14 |
| Number reporting (unspecified) theoretical preparation | 24 |

| | |
|---|---|
| Number graduates of college | *d*81 |
| Number graduates of college and normal school | *e*8 |
| Number graduates of college or normal school | *f*7 |
| Number graduates of normal school only | *f*49 |
| | |
| Number not graduates of either | 28 |
| Number not graduates of college (expressed or implied) | 67 |
| Number failing to report on graduation who reported other data in this table and probably not graduate of either | 23 |
| Total of last three items | 118 |

*a* Three reports specified agriculture in summer school.
*b* This science in some cases may be normal school science, where the report did not indicate whether graduation was from college or from normal school. It includes only the cases reporting science but no agriculture. Most of the agriculture group just above specifically reported science as well.
*c* All but two of these cases were reported from Missouri.
*d* Includes eight or nine graduates of agricultural colleges.
*e* Included in graduates of colleges.
*f* Included in the next item, graduates from normal schools only. These persons answered only "yes" to the question, "Is the instructor a college or a normal school graduate? Which?" In addition to these, five reported as being juniors in college, and five as being juniors in normal school.

Out of the 182 schools noted, 121 are in the three states of Missouri, Nebraska, and Ohio. The principal items of the preceding table relating to those states are distributed as follows:

TABLE 37

PREPARATION OF 121 TEACHERS IN MISSOURI, NEBRASKA, AND OHIO

| | Missouri | Per cent | Nebraska | Per cent | Ohio | Per cent | No. reporting | Total | Per cent |
|---|---|---|---|---|---|---|---|---|---|
| Total schools reporting | 32 | *a*18 | 42 | 23 | 47 | 26 | 121 | 182 | 67 |
| Graduates of college | 9 | 28 | *b*18 | 43 | 18 | 38 | 45 | 81 | 57 |
| Graduates of normal schools only (*c*) | 13 | 41 | 13 | 31 | 9 | 20 | 35 | 48 | 73 |
| Not graduates at all | 9 | 28 | 2 | 5 | 14 | 30 | 25 | 28 | 89 |
| Not reporting, probably not college graduates | 1 | 3 | 9 | 21 | 6 | 12 | 16 | 24 | 67 |
| Total number apparently not college graduates | 23 | 72 | 24 | 57 | 29 | 62 | 76 | 91 | 83 |
| Reporting practical farm experience | 12 | 37 | 12 | 29 | 26 | 55 | 50 | 77 | 65 |
| Reporting agricultural college work | *d*3 | 9 | 2 | 5 | 4 | 8 | 9 | 27 | 33 |
| Reporting work in college science (*e*) | 6 | 19 | 12 | 29 | 9 | 20 | 27 | 41 | 66 |

*a* Percentage of the 182.
*b* A large proportion of the Nebraska science teachers are women from cities, so this item runs higher for Nebraska than do those of farm experience and agricultural courses.
*c* See note *f* under preceding table.
*d* One reported course in summer school.
*e* See note *b* under preceding table.

## Salaries Paid the Teachers of Agriculture, Trained and Untrained

In Missouri, Nebraska, and Ohio, information regarding the salaries paid instructors of agriculture is available, in most cases, in the reports of the state department of education. Except in a few notable cases, this information is not available for the country at large outside of the states named. There is some doubt about two cases in Missouri and seventeen in Nebraska, where the subject was taught by an assistant, whose salary was not specifically stated. Since in none of these instances was a larger salary reported on account of ability to teach agriculture, it seemed safe to use the average salary given for the assistants in the respective schools. Nearly every one of the seven "assistants" reported from Ohio, turned out to be officially listed as "high-school principal," and in every case, was the only other teacher besides the superintendent, and the salary was therefore easily found. The tabulations are shown as follows:

Table 38

Salaries of 110 Teachers of Agriculture in Missouri, Nebraska, and Ohio

| Salary | Missouri | Nebraska | Ohio | Total |
|---|---|---|---|---|
| $301– $400 | 0 | 0 | 4 | 4 |
| 401– 500 | 7 | 4 | 3 | 14 |
| 501– 600 | 8 | 13 | 5 | 26 |
| 601– 700 | 8 | 4 | 8 | 20 |
| 701– 800 | 5 | 4 | 9 | 18 |
| 801– 900 | 2 | 3 | 9 | 14 |
| 901–1,000 | 0 | 0 | 5 | 5 |
| 1,001–1,100 | 0 | 2 | 1 | 3 |
| 1,101–1,200 | 1 | 2 | 2 | 5 |
| 1,201–1,300 | 0 | 1 | 0 | 1 |
| | 31 | 33 | 46 | 110 |
| Average | $640 | $703.80 | $733.70 | $698.35 |
| Average deviation | 106 | 173.00 | 170.00 | ...... |
| Median | 612 | 596.00 | 733.00 | 655.00 |

A few teachers, for instance, superintendents, drawing large salaries may raise the average so as to make it seem to show something not warranted by the facts. Thus the one salary of $1,150 raises the average for Missouri from $623 to $640. The five salaries of $1,100 and over in Nebraska raise it from $624 for the lower twenty-eight teachers to $704 for the thirty-three. The true state of affairs is often better shown by the "median" instead of by the average. The median is the point above and below which there are an equal number of the cases. For example, in Missouri just as many teachers of agriculture receive less than $612 as receive more. For Nebraska, the corresponding middle point is less than $600, over $100 less than the average. The "average deviation" shows how much, on an average, the different salaries vary from some central tendency, in this case the average being used. Thus in Missouri the salaries vary on an average about $106 from the $640, and in Nebraska, about $173. The average is not worth much unless we know how much the individual instances deviate from it, and how the cases are grouped or "bunched." The average and median are practically the same for Ohio because the grouping is pretty much the same toward both extremes.

Reference to Table 39 will show that the largest number of women teachers in charge of classes in agriculture is in Nebraska, where the median is the lowest, and where the average is pulled up by a few well-paid principals teaching the subject. But in Missouri, with almost as low a salary standard, the subject is nearly always taught by the superintendent or principal who is in every case a man. Full data would pull down this figure still further, for information was not asked of any school credited with less than ten students in agriculture in the 1908 report of the Department of Public Instruction.

It is to be borne in mind, of course, that the higher salaries are paid because the recipients are executive and not because they teach agriculture; for they often do it because none of their teachers can or will. Only one of the 110 teachers noted in Table 38, who received over $700, was paid more on this account. He is a superintendent in Missouri and receives $100 extra because of his agricultural ability. Other teachers receiving over $700, and who are paid more on this account, are the high-school assistants at North Adams, Mich., Dillon, Mont., Tyner, Tenn., and the principals of the public high schools at Calvert, Md., Petersham, Mass., and McNabb, Ill.[1] All of the special schools noted later pay more than they otherwise would. Reference has been made to the sex of the teachers here considered. It may be that the relatively large number of women in Nebraska accounts for the low salaries paid there as compared with Ohio, although the figures for Missouri do not seem to lend strong support to this idea.

TABLE 39

OFFICIAL POSITION AND SEX OF THE TEACHERS OF AGRICULTURE IN MISSOURI, NEBRASKA, AND OHIO

| Sex and position | Missouri | Nebraska | Ohio | Total |
|---|---|---|---|---|
| Male superintendent or principal, certain | 27 | 11 | 41 | 79 |
| Female principal, certain | 1 | *a*7 | 2 | 10 |
| Assistant under male principal, doubtful. | 2 | 6 | *b*2 | 10 |
| Assistant under female principal, doubtful | 0 | 9 | 0 | 9 |
| Unknown, doubtful | 1 | 0 | 1 | 2 |
| | 31 | 33 | 46 | 110 |

*a* Two were among those receiving $10 a month extra because of ability to teach agriculture. See Table 40.
*b* Women assistants.

The following table is important because it is the best indication at present of the value placed by boards of education on ability to teach agriculture in the high school. It classifies the 15 out of 184 who are reported as getting more than they otherwise would.

[1] This does not take account of other high schools in the same state into which the subject has been introduced with a special instructor since these statistics were compiled.

TABLE 40

SALARIES OF TEACHERS PAID MORE BECAUSE OF ABILITY TO TEACH AGRICULTURE IN THE HIGH SCHOOL

| Yearly salary | No. | Extra per month | No. | Salary would be less by | No. |
|---|---|---|---|---|---|
| | | | | Per cent | |
| $401–$500 | 1 | 0–$10 | 4 | 11–20 | 8 |
| 501– 600 | 2 | $11– 20 | 4 | 21–30 | 1 |
| 601– 700 | 3 | 21– 30 | 2 | 31–40 | 4 |
| 701– 800 | 4 | 31– 40 | 1 | Over 40 | *a*1 |
| 801– 900 | 3 | 41– 50 | 1 | | |
| Over 900 | *a*2 | 51– 60 | 1 | | |
| | | 81– 90 | *a*1 | | |
| Total | 15 | | 14 | | 14 |

Average of the first 14, $779.20. Average for 14, $25. Average for 14, 22 per cent.

*a* $1,600 was paid for one year to an agriculturist with teaching experience to organize a county high school with agriculture as the prominent feature; but his successors have received probably less than half that amount.

The foregoing table includes only those reported to be getting more on account of teaching agriculture, whether they happened to be trained agriculturists or not. Perhaps a better idea of the market value of the agriculture teacher may be gained by considering the absolute salaries paid the men trained for this work, in the public high school, and in others approximating the type. The following table is based on data from 7 of the congressional district schools of Alabama, 3 of the new state agricultural and normal training schools of Virginia, the 4 county agricultural schools of Wisconsin, 15 county high schools, 4 township high schools, 6 village high schools, and 1 state school in California.

The Alabama, Virginia, and Wisconsin special schools are each represented by one man, the only distinctive agriculturists in the schools, which differ in this respect from the Georgia, Minnesota, and California special schools. In the Wisconsin schools, special teachers of domestic science and of manual training receive from $800 to $1,200.

TABLE 41

SALARIES OF AGRICULTURISTS IN PUBLIC HIGH SCHOOLS AND OTHER PUBLIC SECONDARY SCHOOLS

| Salary | No. |
|---|---|
| $583 | 1 |
| 675 | 3 |
| 750–800 | 6 |
| 900 | 2 |
| 1,000 | 12 |
| 1,200 | 6 |
| 1,300 | 1 |
| 1,450 | 1 |
| 1,500 | 3 |
| 1,700 | 1 |
| 1,800 | 2 |
| 2,000 | 4 |
| | 41 |

All of the first ten are assistants.

Four of the twelve $1,000 men are assistants, one is a principal, and the other seven probably are.

All the others are principals except the one receiving $1,450, and one each of those receiving $1,500 and $1,800.

See also the note under Table 40.

## THE AGRICULTURAL COLLEGE AS A SOURCE OF SUPPLY, WITH STATISTICS ON SALARIES COMMANDED BY RECENT GRADUATES

Passing from the secondary school demand, and its valuation of such service as mentioned, let us now turn to the available supply of persons receiving the most desirable preparation and see what their talents command in other markets.

Table 42 shows what salaries the men just out of agricultural college are receiving when they are engaged by the colleges themselves, by the experiment stations, and for similar work by state or federal departments. The secondary schools must not only compete with the financial inducements here listed, but also with the greater desirability, to the average college man, of college and research work over public school teaching. It must be remembered, too, that many of these graduates are without any teaching experience.

It will be seen from the table that the salaries run quite evenly for the three years, as shown by the comparative average of $947.50, $921.50, and $935.53 respectively. The average deviation being $192, $200, and $190 respectively. The medians are but little different from the averages, viz., $950, $920, and $950. The "mode" or place where most of the cases lie is seen to be very wide and not "bunched," for four-fifths of the salaries are rather evenly scattered from $720 to $1,200.

The presence of 29 men taking higher degrees in the groups receiving the high salaries will, of course, leave a lower average

TABLE 42

DISTRIBUTION OF SALARIES OF GRADUATES FROM AGRICULTURAL COLLEGES IN 1907, 1908, AND 1909

| Salary | All degrees | | | | Higher degrees | | | |
|---|---|---|---|---|---|---|---|---|
| | 1907 | 1908 | 1909 | Total | 1907 | 1908 | 1909 | Total |
| $400– $449 (*a*).. | 0 | 0 | 1 | 1 | .. | .. | .. | .. |
| 450– 549..... | 1 | 2 | 0 | 3 | .. | .. | .. | .. |
| 550– 649..... | 6 | 9 | 3 | 18 | .. | .. | .. | .. |
| 650– 749..... | 16 | 13 | 14 | 43 | .. | .. | .. | .. |
| 760– 849..... | 15 | 18 | 15 | 48 | .. | .. | .. | .. |
| 850– 949..... | 13 | 7 | 17 | 37 | 1 | .. | .. | 1 |
| 950–1,049..... | 16 | 18 | 20 | 54 | 4 | 3 | .. | 7 |
| 1,050–1,149..... | 1 | 1 | 0 | 2 | 0 | 0 | .. | 0 |
| 1,150–1,249..... | 13 | 16 | 16 | 45 | 4 | 5 | 2 | 11 |
| 1,250–1,349..... | 0 | 0 | 0 | 0 | 0 | 0 | 0 | 0 |
| 1,350–1,449..... | 4 | 2 | 5 | 11 | 1 | 2 | 1 | 4 |
| 1,450–1,549..... | 3 | 2 | 1 | 6 | 3 | 0 | 1 | 4 |
| 1,550–1,649..... | 1 | 0 | 0 | 1 | 1 | 0 | 0 | 1 |
| 1,650–1,749..... | 1 | 1 | 0 | 2 | .. | .. | .. | .. |
| 1,750–1,849..... | 0 | 0 | 1 | 1 | .. | .. | .. | .. |
| Total....... | 90 | 89 | 93 | 272 | .. | .. | .. | 28 |

*a* The limits of the groups are placed at the 50's instead of at the 100's, as seven-ninths of the salaries are even hundreds and thus lie at the center of each group instead of at one extreme as would otherwise be the case, making the groups lop-sided.

for the men receiving the bachelor's degree. These 29 men received on an average $1,180 each, with a strong mode at $1,200. The 243 men with the bachelor's degree averaged $909.13, the two receiving $1,700 and $1,800 bringing the average up from $899.

The positions obtained by these 253 men are as follows: Two were elected to professorships, and twelve to adjunct, associate, or assistant professorships. One went into a Louisiana sugar school, one became principal of a township high school,[2] another principal of a county agricultural high school, and a third principal of a school of forestry. Seven went into agricultural commercial work, six into the state, and eight into the federal government service. Fifty-one were elected instructors in colleges, and one hundred and thirty-eight were elected assistants. In many cases it is clear that these are minor officers of administration or of instruction, such as laboratory assistants, etc.; but in other cases it is not clear that they are not on the staff of the experiment station. The twenty-five unclassified men are field agents, experts, or are clearly attached to the experiment stations.

Any one at all conversant with the pittances paid the graduates of classical and literary college courses can readily see the much greater opportunities open to the graduate of the agricultural college.

The following items indicate that the "plums" do not all go to the men taking higher degrees, and the possession of these degrees does not necessarily assure the best-paying or most desirable positions; but the figures do show the handicap on public schools desiring even the bachelor graduates of agricultural colleges.

(1) The three best-paying positions, $1,700 in 1907, $1,680 in 1908, and $1,800 in 1909, went to men holding only the bachelor's degree.

(2) Of the fourteen elected to professorships, either full, adjunct, or assistant, but one was a doctor of philosophy, one a master of arts, one a master of science, and two were doctors of veterinary medicine, a degree which probably does not presuppose the bachelor's degree. These positions pay from $1,100 to $1,600, averaging $1,380. The doctors of philosophy did not fare much better than the others. The four received respectively $1,000, $1,200, $1,200, and $1,600, while the five doctors of veterinary medicine received $1,000, $1,000, $1,200, $1,400, and $1,500. The eighteen holders of master's degrees averaged $1,205, ranging from $900 to $1,500.

[2] A 1909 bachelor of science in agriculture man has just been elected to a normal school professorship.

The eight men entering the service of the United States Government averaged $1,230, ranging from $840 to $1,400.

It must be remembered, however, that some of these men, while graduating from land-grant colleges may have taken general science courses and not courses leading to the degree of bachelor of science in agriculture. It would seem that many of them have done so.

Inquiry made of agricultural college officials revealed the names of very few graduates who had entered public school work. Replies from most of the forty-two whose addresses were obtained showed that only four were then teaching and that not more than three or four others had taught. These college officials and graduates were asked to estimate roughly the salaries high schools must offer to attract the services of agricultural college graduates as principals or as science teachers. Twenty of the former and fifteen of the latter responded. For principalships the estimates were $700 to $1,600 and $600 to $1,500 respectively; for assistantships, $600 to $1,200 and $500 to $1,200 respectively. It will be noticed that there was little difference in the estimates, but that the range is so wide as to make them of little value to a school board in doubt about the probable cost of such services. No distinction could be made between the northern and southern states; but the eastern states, as a group, showed by far the lowest estimates.

## Provisions for the Higher Training of Teachers of Agriculture

Many advances have been made in the way of preparing teachers to meet the new demands for agricultural instruction since Dean Bailey's report[3] of two years ago. Since then the number of colleges providing teachers' courses of from one to four years in length has at least doubled.

While the courses provided for in the regular college curriculum must be depended upon to furnish a substantial foundation for the teacher's preparation in the long run, a more important movement, from the standpoint of immediate results, has been

[3] L. H. Bailey, Bulletin 380, Bureau of Education, On the Training of Persons to Teach Agriculture in the Public Schools, 1908.

the development of the college summer school courses in agriculture, a movement so new that Dean Bailey gives but a passing mention to one institution of college rank offering such work.

To the student of education, probably the most significant development in this field has been the spread of the idea, within the last two years, of establishing chairs of education in agricultural colleges, or chairs of agriculture in colleges of education.

Another movement, yet in its infancy but of great promise, is the idea of the "conference on agricultural education" held in connection with the summer schools. This is a sort of specialized but elaborated teachers' institute of state-wide proportions, and is calculated to focus attention on large professional problems too broad to fit into the class room discussions of special courses.

Agricultural colleges make provision for prospective teachers in one of two ways. Students of the regular four-year course in agriculture may elect courses in education given in the agricultural college or in some other college of the university. The second way is for the prospective teacher to pursue a special group of subjects, supposed to be especially adapted to the needs of teachers in content or in organization. This special group may require anywhere from one to four years to complete according to the number of units it contains. It may not and sometimes does not include any pedagogical work.

The state agricultural colleges of Massachusetts, Michigan, North Dakota, and Indiana (Purdue University) use the first plan. The agricultural colleges of Connecticut and North Carolina, and of the University of Maine have the special group of agricultural subjects without courses in education. Michigan Agricultural College has such a course of one year, which does, however, presuppose a normal school certificate or experience in teaching.

The state universities having agricultural colleges and also colleges of education or well defined departments of education may effect a reciprocal arrangement whereby the latter may furnish the agricultural college students with facilities for pedagogical training. The state universities of Illinois, Minnesota, and Wisconsin, are notable examples of institutions following this plan. Conversely, students in the college of arts and sciences may elect courses in the agricultural college. While many of the courses are technical, some are designed especially for prospective teachers of the subject. The teachers' college of the University of Missouri offers similar agricultural courses for teachers. The " course for teachers " is becoming less and less (except in summer schools) one of subject matter and increasingly one of methods, as the former kind of course is supplanted by groups of more or less technical courses in agriculture.

It is highly significant that nearly all the high-school teachers who reported having received some training in agriculture, other than that gained by practical experience, took courses in summer schools. They were about equally divided between summer sessions of colleges and normal schools. The efficiency of the high-school courses in agriculture of one year or less will doubtless depend for some years upon help the summer schools are able to give science teachers, high-school principals, and village superintendents.

The efficiency of these summer courses is probably increasing more rapidly than their enrollment. This was practically at a standstill in 1908 and 1909 in the twenty-one institutions in United States and Canada most prominent in this work, being about 1,135 each summer. The sessions lasted from two to nine weeks. Five continued four weeks and six ran six weeks each. The registration in agriculture varied in 1908 from 7 to 166. In 1909 the number of summer schools registering less than 25 students fell from eight to four. Over one-third of the students were registered in courses in which the nature-study idea, as commonly understood, was very prominent. About one-half were enrolled in separate agricultural colleges, about one-fourth in the agricultural colleges of state universities, and the rest in the summer sessions of colleges of education or of arts and sciences.

The number of distinct courses along agricultural and nature-study lines offered in each school varied in number from 1 to 19, and were often only one-half or one-third the length of the entire session. In 1908 but three summer schools offered courses restricted to high-school teachers. The exact number of the teachers enrolled in these courses is not available. The following summer four out of six schools offering courses in secondary school agriculture registered 78 persons in these courses. Only the separate agricultural colleges have introduced into their summer work such specialized courses as agricultural pedagogy and rural sociology.

The "conference" is a feature of recent origin in the colleges giving agricultural instruction. It has taken a form, so far, rather distinctive for each institution adopting it. The central topic may be agriculture as a science, as a subject of instruction, or as related to community life.

The conferences held since 1908 during the summer session of the Massachusetts Agricultural College have addressed themselves principally to the first theme and to agricultural science in relation both to the high school and, to a lesser extent, to the elementary school.

During the same years the University of Virginia has held a "rural life week," in which the discussions center more around the improvement of the social and economic condition of agricultural workers.

During the four weeks' session of the North Dakota Agricultural College, at least one lecture of a general nature was given every afternoon. During the "county superintendents' week," an additional series of lectures was given by F. W. Howe, then of the United States Department of Agriculture, now state supervisor of agricultural education for the New York Education Department.

At the University of Illinois, a two days' "conference on the teaching of agriculture in the common schools" was held late in March, 1910, at which twenty-five short papers were given The subject was considered from the standpoint of the university, the normal school, the county superintendent, the school director, the practical farmer, and the woman in the home. A committee was appointed to report on a plan for an eight-grade course of instruction in agriculture at a conference to be held the following winter while the "short courses" were in session at the agricultural college.

The connection between the subject matter of agriculture and the public school is made, as suggested, in two ways. One by a combination of technical courses in agriculture and education; the other by courses in agricultural subject matter, organized with special reference to the needs of public school teachers, and including as much incidental pedagogy as the fitness or inclination of the instructor may permit.

In universities having technical courses in both agriculture and education, the question is one chiefly of administration. The separate agricultural colleges have begun to solve the problem by creating departments to present some of the fundamental work in education. This has been done by the state agricultural colleges of Indiana (Purdue University), Kansas, Massachusetts, Michigan, Mississippi, North Dakota, and doubtless others. The departmental instruction has, so far, generally been given by men trained in the history and philosophy of education and with more or less experience in school administration. More often than not, their special training in agriculture or natural science has been slight.

The second plan is now followed less exclusively than formerly in several state universities, which are effecting a closer union between the technical courses in agriculture and education, and is now being used rather to supplement that combination. Good instances of this change are furnished by the state universities of Illinois, Missouri, and Wisconsin. All of these, as well as the Massachusetts Agricultural College, offer courses usually designated as "agricultural education." Oklahoma Agricultural College has, and University of Tennessee had, an official whose function is, in a way, similar to that of the newly appointed supervisor of agricultural education in the New York Education Department. In most of these cases, the amount of pedagogical training has probably been commensurate with the scientific attainments of the professors of education in the separate state agricultural colleges.

It will be seen then that the term "agricultural education" is used in widely different senses; in some cases meaning principles of education when taught in a college of agriculture, and in other cases meaning principles of agriculture when taught in a college of education. These two kinds of work should be differentiated by being called respectively, "principles of education" and "public school agriculture," or their equivalents. There is also room for an intermediate treatment with a broader outlook than either, which might be called "philosophy of agricultural education" with its complement "methods of agricultural education." The short and expressive term of "agricultural education" more properly belongs to such a synthetic treatment than to work of college grade usually passing by that name.

# CHAPTER VI

# SPECIAL SECONDARY SCHOOLS OF AGRICULTURE

It is the opinion of the commission that there is a demand in various agricultural sections for schools which shall be devoted to specialized work, object lessons, and such practical courses as have a direct bearing on farm life for both boys and girls.—MASSACHUSETTS COMMISSION ON INDUSTRIAL EDUCATION.

No attempt is here made to give a complete account of the technical secondary schools of agriculture, but to draw such a sketch of the schools of this class as will make clear the distinction between their organization and work and that of the general public high school, and show the variety existing among the special schools themselves.

Sample courses of study and something of the equipment of the schools are here given, also some facts about the educational preparation of the principals, the size of the faculties, and the salaries of the agriculturists. Some interesting data are included regarding the students, the extent of their preparation on entering, the number who live at home and travel back and forth every day, and the number of those who live away from home, either boarding in the school dormitories or in homes in the town where the school is located.

## TYPES OF SPECIAL SCHOOLS AND STATE AID

The special agricultural schools have arisen under a variety of legislative enactments. The size of the district in which they have been or may be established also shows a wide variation. Alabama and Georgia have adopted the congressional district as the unit, giving them respectively nine and eleven such schools. Oklahoma has adopted the supreme court judicial district as the unit, and has established a school in each of the five districts, and an additional one in the "Panhandle." The judicial districts average about fifteen counties each.

Arkansas was divided by its legislature during the past year 1909 into four agricultural school districts, with from seventeen to twenty counties in each, in each of which a school has now been located. The county has been adopted as the unit by Michigan,[1] Minnesota, Mississippi, and Wisconsin. The first four leave the establishment of such a school optional with any county, while the number that may be established in Wisconsin is limited. Two were permitted by the law of 1901; this number was increased to four in 1903, and to eight in 1907. One school of this type has been established in Michigan, but none has been in Minnesota. Minnesota has, however, established as a branch of the state university a "state agricultural high school" at Crookston, which is not supported by any particular district and is supposed to minister to the northwestern section of the state in general.

Other state schools established as branches under the control of the state agricultural college are those at Davis, Cal., and Dahlonega, Ga. Among independent state schools are the California Polytechnic School, at San Luis Obispo, Cal.; the College of Industrial Arts (for women) at Denton, Tex.; and three schools in New York, two established in connection with colleges, one at Canton, in 1906, and one at Alfred in 1908, and one separate school established the latter year at Morrisville.

New York, by an act which became a law April 22, 1910,[2] amended a law passed in 1908, broadening the powers of cities and union free school districts, enabling them to establish "schools of agriculture, mechanic arts and home making" as well as "general industrial schools and trade schools." The authorities are given full powers to provide competent teachers, curricula, ground, buildings, and necessary supplies. The state grants "to each city and union free school district the sum of $500 for each independently organized general industrial school, trade school, or a school of agriculture, mechanic arts and home making, maintained therein for thirty-eight weeks during the school year and employing one teacher whose work is devoted

[1] But there is no prospect of the early establishment of such schools in other counties than Menominee.

[2] Laws of New York, chap. 140, art. 22, sec. 600–607.

exclusively to such school, and having an enrollment of at least twenty-five pupils . . . and an additional $200 for each additional teacher employed exclusively in such schools for thirty-eight weeks during the school year."

Massachusetts gives state aid to a school established by private benefaction at Northampton; and its industrial commission is trying the plan of converting certain high schools into agricultural schools, when other schools in the same townships are equipped to give instruction in the classics. An example of this movement is seen in the course of study being put into operation at the Montague High School, now being attended by pupils from five neighboring towns.

## Support and Control

In Alabama the "board of control" consists of the governor, the superintendent of education, the commissioner of agriculture, a secretary-treasurer, a resident member, and one other member selected from the district. The amount of state support has risen from $2,500 given each of the two schools originally established, in 1889, to $4,500 at the present time.[3] Each school has an experimental farm in its vicinity in charge of a trained agriculturist. In three of the schools, the experiment station and the instruction in agriculture are in charge of the principal. The law requires that $750 of the state appropriation shall be expended on the experiment station. The printed course of study shows agriculture now required in all four years. Fees for tuition, library, or incidentals, range from nothing up to $12, with books bought by the students at an average annual cost of a little more than $7. Some schools charge a matriculation fee. The location of these schools is shown in the Table 43.

The bill providing for the establishment of the eleven district agricultural schools of Georgia, passed in 1906,[4] provides that "they shall be branches of the State College of Agriculture,

[3] C. J. Owen, Bulletin 220, Office of Experiment Stations, 1909. A full account is here given of the history of the legislation regarding the establishment, from time to time, of the other seven schools, and the support given to all the district schools.

[4] Georgia Statutes, Act. No. 448, p. 72, Aug. 18, 1906.

a department of the University of Georgia," and that "the general board of trustees of the university shall exercise such supervision as in their judgment may be necessary to secure unity of plan and efficiency in said schools."

The local boards of trustees consist of one member from each county of the respective congressional districts, appointed by the governor for a term of six years. The schools receive the income from the fertilizer, oil, and other inspection fees, over and above the expense of such inspections. This amounted in 1909 to $7,250 annually for each school. In addition to this amount the legislature voted a grant of $2,000 to each school, making the total $9,250. The students are charged no fees for tuition, laboratory, library privileges or incidentals. They buy their own books at an average cost of $7.50 a year. Their board amounts to $10 a month, with a rebate of $5 for farm work performed according to the requirements of the course of study. The local communities provide the necesary land, buildings, and equipment. The location of these schools is shown in Table 44.

The Oklahoma schools of agriculture are under the general management of a "state commission of agricultural and industrial education," consisting of the state superintendent of public instruction, the president of the state board of agriculture, and the president of the Oklahoma Agricultural and Mechanical College. The state board of agriculture exercises a general oversight of these schools, while their work is under the more immediate direction of a dean of the department of district agricultural schools, attached to the college. A condition of the location of the schools is that they "shall be provided with not less than 80 acres of land without cost to the state and deeded in perpetuity to the state. All white citizens over fifteen years of age are entitled to admission without entrance examination or fees." As a consequence of this statutory provision the schools carried on sixth, seventh, and eighth grade work, during the first year with as many as thirty in a class. The work of the secondary grade extends over three years, and offers nothing besides the purely industrial courses and related science, except in mathematics, English, civics, and history. One-fourth of the $20,000 appropriated for building and maintenance the first year, was to be expended for "agricultural experiment in field, barn, orchard, shops, and garden."[5]

Each school in Arkansas is controlled by a board of five trustees "who shall be intelligent farmers," appointed by the governor for a term of ten years. They may fix the rules of admission so as to equalize the attendance among the counties. They may limit the number to suit the capacity of the school but shall not charge tuition. Students must be fifteen years of age. An initial appropriation of $40,000 was made for each of the four schools to supplement donations from the local communities. The law requires that after the first buildings are erected and ready for temporary use all work connected with the care and operation of buildings, farm, stock, etc., shall be performed by the students.

Another interesting feature is the following provision: ". . . not later than one year succeeding the opening of each school, there shall be established in connection therewith a textile school in which shall be taught the art of cotton manufacturing, and other manufacturing should the board of trustees deem it expedient."[6]

The location of these schools is as follows: Jonesboro, in the northeastern part of the state, for the first district, Russellville in the northwestern part for the second district, Magnolia, in the southwestern part, for the third district, and Monticello in the southeastern part, for the fourth district.

In Michigan, county agricultural schools may be established by single counties, or by two or more counties jointly. The appropriations must be made by the county supervisors. They also elect the four members of a county school board, who, with the county school commissioner, have charge of the operation of the school. Where two or more counties unite to maintain such a school each county furnishes two members of the agricultural school board, of which the school commissioner of the county in which the school is located is also a member. On determining to establish such a school by a two-thirds vote of its members, the county supervisors must submit the proposi-

[5] Chap. 3, Senate bill 109, p. 13, May 20, 1908, Oklahoma legislature.
[6] House bill No. 2, session of 1909, Arkansas general assembly.

tion to the voters before issuing bonds or contracting any indebtedness. The only school so far established under the provisions of this act of 1907,[7] is the one at Menomonie, started the next year. It is across the river from Marinette, Wis., where is located a county agricultural school which opened in the year of this legislation.

The notable experiment in secondary agricultural education started by Wisconsin resulted from the report and recommendations of the state superintendent, L. D. Harvey, to the legislature of 1901 after an extended investigation. The result was the authorization of the two schools that started the following year, one at Menomonie, Dunn County, and the other at Wausau, Marathon County. Each school is controlled by a county board of three members and is under the general supervision of the state superintendent, who "with the advice of the dean of the college of agriculture of the State University shall prescribe the courses of study to be pursued and determine the qualifications required of the teachers employed in such schools." The original provision was that the state should bear one-half of the annual instructional expense of each school, provided that not more than $2,500 should be so paid. In 1903 the law was amended so that two-thirds of the annual cost of maintenance of each school should be paid out of the state treasury, with the maximum limit placed at $4,000, and provided that any deficit should be paid by the county. That the counties themselves do not pursue a niggardly policy, is evident from the annual statements, which show that they not only provide their full share of $2,000 but often several hundred more each year. The establishment of the school at Winneconne, Winnebago County, two months after the Marinette county school, filled up the number authorized by the legislature in 1903. In 1907 the maximum number was increased to eight.[8] This additional allowance will soon be exhausted as LaCrosse County has already established a school at Onalaska, while Brown and Langlade counties have voted to establish similar schools. Vil-

[7] Act No. 35, April 3, 1907, [Michigan] public laws.

[8] Wisconsin, Laws of 1901, chap. 188, sec. 10; Laws of 1903, chap. 143, (sec. 5531, Statutes); Laws of 1907, chap. 540, July 10.

lages are also authorized to issue bonds to bear part of the expense of county schools, not to exceed one-fifth of the cost of the school.[9]

No county agricultural high schools have been established in Minnesota, although counties are permitted by the legislation of 1905 to appropriate as much as $20,000 in one year to establish and maintain schools of agriculture and domestic economy. Two or more counties may unite for this purpose. The county school board is composed of the county school superintendent, who acts as secretary, and two other members chosen by the county commissioners. The relation of the school to the state department of public instruction and to the state college of agriculture is the same as in Wisconsin. Tuition is free to residents of the supporting counties.

The legislature of Mississippi, in 1908, enacted a law providing that "it shall be lawful for the county school board of any county to establish one agricultural school in the county for the purpose of instructing the white youth of the county in high school branches, theoretical and practical agriculture, and such other branches as the board hereinafter provided for may make a part of the curriculum."[10] The "board of trustees" was to be composed of five members, the county superintendent of education, two members elected by the board of supervisors, and two by the county school board, to serve four years. The annual tax levy was not to exceed two mills, and state aid was limited to $1,000 for any one county.

The supreme court declared the law unconstitutional because it did not make equal provision for the establishment of like schools for the colored youth. A new act was approved March 16, 1910, designed to remedy this defect. By its provisions, the "board of trustees" is constituted as before, the limit of tax levy is the same, but the maximum of state aid is raised to $1,500 for any one county, or $3,000 for two counties maintaining joint schools. The two counties may unite in maintaining one school for each race. The joint board is composed of eleven members, five chosen from each county as provided in

[9] Wisconsin, Laws of 1907, chap. 11, Mar. 16.
[10] Laws of the State of Mississippi, 1908, chap. 102, pp. 92–93, Mar. 21.

the case of separate county boards, with an additional member chosen by the two boards, or by lot from the two highest in the voting. The grant from the state treasury is optional with the state board of education, which may not inspect the school until certification has been made that the school is provided with twenty acres of land and suitable buildings, including a dormitory accommodating at least forty persons.

Section 1 of the recent act provides for the establishment of "not more than two agricultural high schools in the county, one for white youths exclusively and the other for colored youths exclusively," and adds that "if only one school is established at first, the school board shall have the power at any subsequent time to establish an additional school whenever the necessity for the same shall arise."

Section 2 prescribes the manner of levying and collecting the tax, and provides "further, that within twenty days after a levy has been made twenty per cent of the qualified electors of the said county may file with the clerk of the board of supervisors a petition asking that the tax for the support of either one or both agricultural high schools be not levied, then the questions shall be submitted to an election of the qualified electors of the county within thirty days after the next meeting of the board of supervisors after the filing of the petition, at which election said electors may vote against the tax levied for the support of either one or both schools; and should a majority of the votes be cast against the tax levied for the support of either one or both schools, then the levy of the board for the support of that school or of both schools, as the case may be, shall be null and void and the tax collector shall refuse to collect such tax so voted against."

Whether this will enable the voters to support the school established for one race, and refuse to permit the establishment of the school for the other race, I do not know.

Six county high schools had been organized before the adverse decision of the supreme court stopped further efforts. They are located as follows: Bay Springs, Jasper County; Buena Vista, Chickasaw County; Camden, Madison County; Kossuth,

Alcorn County; Oakland, Yalobusha County; and Poplarville, Pearl River County.

The fact that both Latin and Greek are taught in some of these schools makes it of doubtful propriety to include them with agricultural high schools as strictly defined.

Reference has been made in Chapter II to schools of the public high-school type which add agriculture to their curricula and then appropriate the name "agricultural high school." While in some cases such schools may be regarded merely as variants of the usual high school with a diversified curriculum, it is hardly fair to so regard certain other schools which are really a sort of cross between the general high school and the one which is exclusively industrial. The fact that the state makes special grants to some high schools to establish definite departments of agriculture, and that the central authorities organize the work instead of leaving its initiation and organization entirely to local effort, would seem to place these schools in a class by themselves. But even here we find a very close analogy in the procedure of such states, as New Jersey, which give state aid to schools maintaining a department of "manual training," or "industrial arts," depending in amount on the money spent by the local board. In no case, however, do we find these schools calling themselves manual training or industrial schools, when they carry on all the other work usually taught in high schools. Schools are rapidly being reorganized on this plan in Louisiana, Minnesota, and Virginia.

To many thoughtful educators, especially to those in sympathy with the views of Dean Davenport,[11] of the University of Illinois, this intermediate type of school may seem more desirable than the Wisconsin type, combining the strong points of the latter with those of the general high school, and avoiding the disadvantages of both.

State secondary schools of agriculture other than those organized in connection with the agricultural colleges and using their equipment, have been established in but three states, as already mentioned. These are more or less dependent upon such appro-

[11] Education for Efficiency, 1909, chap. V and VI.

priations from year to year as the legislatures may be prevailed upon to grant. Such legislative appropriations are apt to be larger to these individual schools than is the share given each of the several schools of a group or class, all established within a short period.

The California Polytechnic School, at San Luis Obispo, is governed by a board of seven trustees, of which the governor and state superintendent of public instruction are ex officio members. While about half the students come from San Luis Obispo County, the central and southwestern counties are rather generally represented. The school was established in 1901 by act of the legislature, which has liberally supported it and has enabled it to add to its holdings and buildings as its growth demanded. For instance in 1907, about $80,000 was voted for improvements over and above the running expenses. While there are no tuition fees, laboratory fees of $15 are charged for all courses. Text-books and supplies cost about $15 more.

The state university supports a secondary school of agriculture at the University Farm at Davis, started in 1908. For its purchase and equipment, the legislature appropriated $132,000 the previous year.

The University of Minnesota has also established a branch agricultural school some distance from its college of agriculture. The Crookston School of Agriculture, as it is known, is theoretically a high-school department of the college of agriculture, ranking with the "State Agricultural High School," as it is called, which is maintained at St. Anthony Park, near the college of agriculture at St. Paul. The school at Crookston supposedly gives the same course as is given at the parent secondary school with such variations as local demands necessitate. The first superintendent regarded it as the special school for the Ninth Congressional District, including the Minnesota side of the Red River Valley, although the school is open to residents of the state generally.[12] The Crookston school is under the control of the regents of the university as represented in the management of the college of agriculture. Its budget seems to be separate, and appropriations are made specifically for it

[12] Wm. Robertson in *Minnesota Farm Review*, Sept., 1908, p. 164.

by the legislature. The first appropriation was made in 1905, $15,000. In 1907, $65,000 was appropriated for a dormitory, dining hall, and an industrial building. The fees are: matriculation, $5, book rent for those not desiring to buy books, $2, depreciation of room furnishings, $2, board and room, $15 a month.

New York established its first "school of agriculture" in July, 1906 at St. Lawrence University, Canton. Eighty thousand dollars was appropriated for a building. While the school is maintained by the state, its affairs are administered by the trustees of the university. The charter under which it operates has been amended to restrict the agricultural instruction to elementary and practical courses.

A like amount was appropriated in 1908 for a similar school at Alfred University, at Alfred. Five thousand dollars of this was available for the first year's maintenance, after which the annual maintenance fund is to be $10,000.

The amount granted the same year for the establishment of the agricultural school at Morrisville was $20,000. In 1910 the state appropriated $59,275, and the title to a group of former county-seat buildings has been transferred to the state. The school opened in October, 1910.

As these are isolated schools that cannot be advantageously grouped for consideration according to a general plan, they will not be treated at length. The schools of Alabama, Georgia, Oklahoma, Wisconsin, and Michigan, however, do admit of comprehensive treatment on the group plan. The description here given will enable the reader to make some comparison of the plan, scope, and work of these types of agricultural schools, representing different local conditions of wealth and culture.

## Organization, Curriculum, and Equipment

In their most complete development, the special agricultural schools have as the most prominent feature of organization, a three-fold division of the industrial work, namely, departments of agriculture, manual training, and home economics, with the academic studies forming a subsidiary part of the school curriculum. The latter studies are designed to prevent the education of the student from proceeding along too narrow lines and to remedy deficiencies in general culture usually quite obvious. Rural high schools, instituting industrial work under special legislative authority, and some of those schools doing so in the absence of special legislation, show a tendency to follow the same plan.

### *The Alabama Congressional District Schools*

The congressional district agricultural schools of Alabama have been greatly benefited by a standardization made possible by the Association of Presidents and Agriculturists of the nine district schools, organized in 1907. One result of their labors is the course of study here given, which went into operation in the fall of 1909. Before that time there had been no uniformity as to the requirement or election of Latin, agriculture, or the amount of time for the latter. Nor were the entrance requirements the same. In 1909 five schools reported that the sixth grade was the highest that all of their students had completed. Other schools reported the requirements then in effect as being as high as the completion of the eighth grade. The course here given is based on seven years of elementary work.

The percentages given below show what proportion the class-room work in agriculture and related sciences is of the total class-room work, and what proportion all the agricultural work, including the two hours a week on the experiment station, is of the total time required of the student in the class room, laboratory, and in the field.

TABLE 43

TIME GIVEN TO AGRICULTURE IN PROPORTION TO OTHER SCHOOL WORK

| | First year | Second year | Third year | | Fourth year |
|---|---|---|---|---|---|
| | | | First term | Second term | |
| | Per cent | Per cent | Per cent | Per cent | Per cent |
| Proportion of class work | 22 | 22 | 9 | 19 | 25 |
| Proportion of all time including farm work | 31 | 31 | 19 | 28 | 32 |

## CURRICULUM OF THE CONGRESSIONAL DISTRICT AGRICULTURAL SCHOOLS OF ALABAMA[13]

[(*a*) and (*b*) placed after a subject indicate first and second semester, respectively.]

| Agricultural-scientific course | | Agricultural-classical course | |
|---|---|---|---|
| Subjects | Periods per week | Subjects | Periods per week |
| | *First Year* | | |
| English grammar and composition | 5 | | |
| Arithmetic | 5 | | |
| Algebra | 5 | Substitute Latin for English history and physical geography | 5 |
| English history | 3 | | |
| Physical geography | 2 | | |
| Elementary agriculture | 3 | | |
| Practical work | 1 | | |
| | *Second Year* | | |
| English | 5 | | |
| Algebra | 5 | | |
| Ancient history | 3 | | |
| Physiology | 4 | Substitute Caesar for physiology | 5 |
| Soils and crops (*a*) | 5 | | |
| Agricultural botany (*b*) | 5 | | |
| Practical work in each | 1 | | |
| | *Third Year* | | |
| English literature | 5 | | |
| Plane geometry | 5 | | |
| Arithmetic reviewed (*a*) | 3 | | |
| Pedagogy (*b*) | 3 | | |
| Mediaeval history (*a*) | 3 | | |
| Modern history (*b*) | 3 | | |
| Physics with laboratory work | 5 | | |
| Stock lectures and agricultural literature (*a*) | 2 | Substitute Cicero for stock lectures and agricultural literature during first semester | 5 |
| Horticulture (*b*) | 2 | | |
| Agricultural botany (*b*) | 3 | | |
| Practical work each term | 1 | | |
| | *Fourth Year* | | |
| American literature | 5 | | |
| Solid geometry (*a*) | 3 | | |
| Plane trigonometry (*b*) | 2 | | |
| American, history, civics | 3 | | |
| Chemistry with laboratory work | 5 | | |
| Agriculture (*a*) dairying, soils, and fertilizers | 5 | (May substitute Virgil for soils, fertilizers and agricultural literature) | 5 |
| (*b*) Agricultural literature or farm accounts | 5 | | |

NOTES.—All students must work two hours a week on the experiment station. Courses are offered the girls in music, expression, and art. One school offers courses in home economics. One school has a complete woodworking outfit, costing $1,500. Two schools have commercial courses.

[13] This program of studies and the next table are adapted from the report of President C. J. Owens, Bulletin No. 220 of the Office of Experiment Stations, with information from the individual school catalogues.

This course gives about 17 periods of, perhaps, 40 minutes each a week, for each of the four years, or 12 hours of class-room work other than agricultural, and about 5 periods, or 3½ hours of industrial work, with 2 hours' work on the experiment station.

It will thus be seen that these schools act more or less as commercial and finishing schools for their respective districts. Physical and chemical laboratories are specifically mentioned for five schools, school gardens on the campus for three, school gardens and experimental plats on farms near the school in two other cases; farm dwellings are mentioned in the case of four schools, live stock for six, and barns, or live stock implying barns, in the case of seven schools. One reports a full poultry outfit, and another reports a dairy. The total value of all plants is $222,500. Literary societies are reported by eight schools, and athletic organizations by four, one of these also reporting a military battalion.

Other items are tabulated below.

TABLE 44

EQUIPMENT AND FACILITIES OF THE CONGRESSIONAL DISTRICT SCHOOLS OF ALABAMA

| District | Year established | Acres in experiment station | Acres in campus | Library volumes | Buildings value | Faculty | |
|---|---|---|---|---|---|---|---|
| | | | | | | Men | Women |
| First, Jackson | 1896 | 49 | .. | 600 | ..... | 4 | 0 |
| Second, Evergreen | 1893 | 50 | 10 | 500 | $12,000 | 2 | 2 |
| Third, Abbeville | 1889 | 80 | .. | 2,000 | 32,000 | 3 | 1 |
| Fourth, Sylacauga | 1897 | 40 | .. | ... | 32,000 | 4 | 1 |
| Fifth, Wetumpka | 1895 | *a*80 | .. | ... | 41,000 | 3 | 3 |
| Sixth, Hamilton | 1895 | *b*40 | 2 | ... | ..... | 3 | 1 |
| Seventh, Albertville | 1894 | 48 | 1 | ... | ..... | 3 | 3 |
| Eighth, Athens | 1889 | 150 | 3 | ... | 12,000 | 3 | 2 |
| Ninth, Blountsville | 1895 | 75 | .. | ... | ..... | 3 | 2 |
| Total | | *c*612 | .. | ... | ..... | 27 | 15 |
| Teachers reported as holding degrees | | | | | | 22 | 6 |

*a* Fifteen acres are devoted to the experiment work.
*b* The number given is the number of acres devoted to the experiment work.
*c* Bulletin of the University of Georgia, Dec., 1909, p. 14, gives a total of 640 acres.

### *The Georgia Congressional District Schools*

The congressional district schools of Georgia[14] act concertedly in many ways. Joint meetings are held by the various district boards of trustees, or their representatives, and by the principals or other representatives of the faculties, to consider matters calling for some uniformity of action.

The board of trustees of the University, in January, 1907, adopted certain resolutions regarding the work of the schools. A condensed statement of some of these is here given: The minimum age of entrance shall be 14 years for boys and 13 for girls. The course of study shall be limited to four years of forty weeks each, including one year of the common branches. The program shall arrange for at least three hours devoted to academic work and at least three hours in the laboratory, shop, or on the farm, with the program so arranged as to provide, by alternation of class and practical work in the morning and afternoon, for the continuous operation of the shop and farm. Satisfactory labor on the farm or in the shop shall be credited to the dormitory account of the students at a fair rate, either by the hour, or by the piece. Their account shall also be credited with the pro rata of the net profits arising from the farm. One-fourth of the students, or as many as the principal shall deem necessary, shall be required to remain during vacation to continue the operation of the farm and shop. The schools are required, so far as practicable, to provide short courses for adults. The state farmers' institute director is expected to conduct institutes at the schools and to use the instructors in the various county institutes.

Eight schools report paying the students at the rate of ten cents an hour and one at seven cents. The schools not allowing credit for student work are two of the three charging low rates for board, so that it is more or less evened up. All schools allow pay for work beyond the nine hours which they may de-

[14] Most of the information regarding matters of administration is from the report of the Board of Trustees of the University of Georgia, issued July, 1907, and the first annual report of the agricultural schools, by Professor J. F. Stewart, of the University, issued December, 1909. The student statistics are partly from the latter, partly from my own returns from the schools, and partly from their catalogues.

mand. Student labor has been used in building barns, engine houses, water towers, and roads, outside the routine work.

## CURRICULUM OF THE CONGRESSIONAL DISTRICT AGRICULTURAL SCHOOLS OF GEORGIA

### *First Year*

| | | |
|---|---|---|
| **Class-room work:** | | |
| English | periods | 4 |
| Arithmetic | do | 4 |
| United States history | do | 3 |
| Geography (a) political, (b) physical, (c) commercial | do | 3 |
| Penmanship and spelling | do | 2 |
| Agricultural science (a) structure and physiology of plants, (b) environment and reproduction of plants, (c) soils, (for boys) | do | 3 |
| **Practicums:** | | |
| Laboratory work with plants. Plat work, gardens (elective for girls) | hours | 3 |
| Farm mechanics (free-hand and mechanical drawing, all) (b and c) drawing, bench and carpentry work (boys) | do | 3 |
| Home economics (a) sewing, (b) cooking, (c) sewing and laundry | do | 3 |
| Class-room work, 19 periods for boys | do | $12\frac{2}{3}$ |
| Class-room work, 16 to 19 periods for girls. | | |
| Laboratory, garden and shop work, for boys | do | 9 |
| Laboratory, garden and shop work, for girls | do | 6 to 9 |
| Minimum farm work for boys, and home work for girls | do | 9 |

### *Second Year*

| | | |
|---|---|---|
| **Class-room work:** | | |
| English | periods | 4 |
| Mathematics (a) farm arithmetic and accounts, (b and c) algebra and farm arithmetic | do | 4 |
| Ancient history | do | 3 |
| Agriculture, forestry and horticulture | do | 2 |
| Agriculture (a) soils and fertilizers, (b and c) farm crops (for boys) | do | 3 |
| **Practicums:** | | |
| Laboratory (a) soil experiments and farm crop systems, (b) plants, judging, grafting, etc., (c) early fruits and vegetables | hours | 3 |
| Plat work (a) gardens, (b) manures and winter crops, (c) gardens and spraying of fruits | do | 3 |
| Farm mechanics (a and b) plants for farm structures, carpentry, (c) farm blacksmithing (for boys) | do | 3 |
| Home economics (a) sewing, (b) cooking, (c) household emergencies | do | 3 |
| Class-room work, 16 periods for boys | do | $10\frac{2}{3}$ |
| Class-room work, 13 periods for girls | do | $8\frac{2}{3}$ |
| Laboratory, garden, and shop work, for boys | do | 9 |
| Laboratory, garden, and shop work, for girls, hours doubtful | do | 6 |
| Minimum farm and home work | do | 9 |

*Third Year*

Class-room work:

| | | |
|---|---|---|
| English | periods | 4 |
| Algebra (fall term) | do | $1\frac{2}{3}$ |
| English history | do | 2 |
| Physics | do | 3 |
| Agriculture (a and b) animal husbandry, (c) dairying | do | 3 |
| Home science (for girls) | do | 3 |

Practicums:

| | | |
|---|---|---|
| Laboratory, agricultural physics, soils, farm and dairy machinery | hours | 3 |
| Field work (a) crops, stock judging, (b) care of stock, breeds of stock, (c) study of farm buildings | do | 3 |
| Mechanics (a) blacksmithing, plumbing, steam fitting, (b) farm machinery, (c) building construction, concrete work. | do | 3 |
| Home economics (a) sewing, millinery, (b) cooking, (c) sewing, hygiene | do | 3 |

Class-room work (a) 16 periods, (b and c) 11 periods.
Class-room work (a) 16 periods, (b and c) $10\frac{2}{3}$ hours and $7\frac{1}{3}$ hours.

| | | |
|---|---|---|
| Laboratory, field, and shop work for boys | do | 9 |
| Minimum farm and home work | do | 9 |

*Fourth Year*

Class-room work:

| | | |
|---|---|---|
| English | periods | 4 |
| Geometry | do | 4 |
| Civics (a) 3 periods, (b) 2 periods | do | 2 |
| Chemistry | do | 3 |
| Agriculture (a) rural engineering, (b) farm management, (c) rural economics | do | 3 |
| Home science | do | 3 |

Practicums:

| | | |
|---|---|---|
| Laboratory (a and b) chemistry of foods, feed-stuffs, fertilizers and animal products, (c) bacteriology | hours | 3 |
| Field work (a and b) surveying, laying out fields, drains | do | 3 |
| Mechanics (a) drawing farm plans, (b) topographic drawing, construction of roads | hours | 3 |
| Library reading (spring term) | do | 3 |
| Home economics (a) sanitation, (b) household decoration, planning and management of a home | do | 3 |

| | | |
|---|---|---|
| Class-room work (a) 17 periods, (b) 16 periods, (c) 14 periods—$11\frac{1}{3}$, $10\frac{2}{3}$, and | do | $9\frac{1}{3}$ |
| Laboratory, field and shop work for boys, fall and winter | do | 9 |
| Laboratory and library reading for boys, spring term | do | 3 |
| Laboratory and library reading for girls, fall and winter | do | 6 |
| Laboratory and library reading for girls, spring term | do | 9 |
| Minimum farm and home work, as before | do | 9 |

A tentative course of study was planned by Mr. D. J. Crosby, Specialist in Agricultural Education in the Department of Agriculture, who outlined the industrial work, and Prof. J. S. Stewart, of the department of secondary education, in the University, who planned the non-industrial part. This plan was referred to a committee, and, on June 19th, was accepted as modified by the committee and recommended to the several local boards.

Modern languages were included in the original plan but were eliminated by the committee. It will be noted that as much mathematics is included as is usually found in secondary schools. These schools must offer it as they prepare for the State Agricultural College. The catalogues of the schools show local variations from the model course of study. Most of the changes are minor, usually changes in relative position. Agricultural arithmetic does not seem particularly agricultural so far as the text would indicate, though texts are written on this basis. The industrial trend is more evident from the texts used in the course in bookkeeping. In some instances plane trigonometry and surveying appear. Standard texts in botany appear to be used as guides for the class-room work in first-year agriculture. This can not be avoided in the present condition of the text-books, although some recent texts in "elementary agriculture" include a large proportion of technical botany. Physical geography occasionally appears as a second-year study, as does also chemistry.

An examination of the course of study given above shows that not only does it provide that a very large proportion of the total time of the student shall be spent in industrial work, but that a very considerable amount of the class-room work is also of this nature, counting agricultural arithmetic and the sciences.

TABLE 45

TIME GIVEN TO AGRICULTURE IN PROPORTION TO OTHER SCHOOL WORK

| | First year | Second year | Third year | Fourth year |
|---|---|---|---|---|
| | Per cent | Per cent | Per cent | Per cent |
| Proportion of class work | 16 | 45 | 40 | 36 |
| Proportion of all time exclusive of farm labor | 50 | 75 | 72 | 77 |

The above percentages are approximate rather than exact. They hold fairly true for the boys, but would vary five to fifteen per cent lower for the girls.

The course averages about 11 periods of 40 minutes each a week for each of the four years, or 7½ hours of class-room work other than agricultural, and about 12½ hours in industrial work, including class work, laboratory, field, or shop practice, with a minimum of 9 hours of farm work.

TABLE 46

DATA ON THE PROPERTY AND INCOMES OF THE GEORGIA CONGRESSIONAL DISTRICT SCHOOLS

| District | Acres of land | Value of land | Cash promised (*a*) | Value of plant | Debt, 1909 | Income, 1909 (*c*) | Pupils fall term, 1909 |
|---|---|---|---|---|---|---|---|
| First, Statesboro | 300 | $15,000 | $60,000 | $100,000 | $2,500 | $11,500 | 56 |
| Second, Tifton | 315 | 15,000 | 60,000 | 65,000 | 2,000 | 12,750 | 50 |
| Third, Americus | 270 | 15,000 | 40,000 | 70,000 | ..... | 10,250 | 88 |
| Fourth, Carrolton | 275 | 13,000 | 30,000 | 60,000 | 4,000 | 13,850 | 110 |
| Fifth, Monroe | 250 | 8,000 | 31,000 | 60,000 | 10,000 | 12,650 | *e*74 |
| Sixth, Barnesville | 312 | 13,000 | 51,000 | 100,000 | 4,000 | *d*12,500 | *e*43 |
| Seventh, Powder Springs | 240 | 8,400 | *b*25,000 | 40,000 | ..... | *a*11,750 | 66 |
| Eighth, Madison | 257 | 20,000 | 40,000 | 60,000 | ..... | 10,750 | 22 |
| Ninth, Clarksville | 300 | 5,000 | 25,000 | 45,000 | 6,500 | *a*9,450 | *e*83 |
| Tenth, Granite Hill | 265 | 8,100 | 47,000 | 52,000 | 3,500 | *a*13,420 | 58 |
| Eleventh, Douglas | 300 | 15,000 | 55,000 | 90,000 | 4,000 | 9,250 | 100 |
| Total | 3,084 | $135,500 | $464,000 | $742,000 | $36,500 | $128,120 | 750 |

*a* All but the seventh, ninth, and tenth districts promised free electric light, water for five years, and sewage disposal.
*b* Promised an academic building and boys' dormitory.
*c* Includes farm profits reported by all but one school, gift to one school, and profits from boarding department reported by five, as well as the uniform sum of $9,250 received from the state.
*d* Approximate, as boarding profits are reported for only fall term.
*e* Fifth, sixth, and ninth district schools are one year below the other schools in grade.

The students cultivated 738 acres of the total acreage of 3,084. Only two schools employed outside farm help, one man each during the year. Two schools rented 45 and 55 acres respectively. One hundred and seventy acres were put in cotton, 415 in grain, and 543 in pasturage and other crops. The value of the farm crops for 1909 was $16,050. The eleven schools possess 16 barns, 32 mules, 8 horses (five schools), 31 sheep and goats (three schools), 255 hogs, 855 fowls, and the following equipment:

Farm implements $3,550, ranging from $50 to $800.

Dairy equipment $935, ranging from $50 to $200.

Shop equipment, $2,990, ranging from $60 to $900.

Laundry equipment, $4,090, ranging from $15 to $2,000 (none in three schools).

Domestic science, $1,715, ranging from $50 to $600.

One can not help being somewhat surprised at first thought, at the relatively small amount of money spent upon the laboratories, including even the agricultural laboratory facilities. Five report none, while the remaining six report a total of only $275. Six report no chemical equipment, the remaining five report a total of $505. Only three schools report physics, with a total equipment of $385. The last two items are not surprising, however, standing alone, for the schools have not been running long enough to put into operation that part of their course of study in which these branches occur. Then, too, it is often hard to draw the line between apparatus belonging to agriculture and to other sciences.

Eight schools report 1,920 volumes, ranging from 10 to 600 to the library. The total value is placed at $975, for six libraries varying from $25 to $400 each.

## *The Oklahoma District Schools*

One of the Oklahoma schools was in operation during the year 1908-9, the Murray State School of Agriculture, at Tish-

TABLE 47

DATA ON THE PROPERTY AND INCOMES OF THE JUDICIAL DISTRICT SCHOOLS OF OKLAHOMA

| District | Acres of land | Income for year ending— | | Faculty | |
|---|---|---|---|---|---|
| | | June 30, 1910 | June 30, 1911 | Men | Women |
| First, Warner | 160 | $17,000 | $14,000 | 4 | 1 |
| Second, Tishomingo | 100 | 17,000 | 14,000 | 4 | 1 |
| Third, Broken Arrow | 120 | 12,000 | 17,000 | 4 | 1 |
| Fourth, Lawton | 80 | 12,000 | 17,000 | 4 | 1 |
| Fifth, Helena | 100 | 12,000 | 17,000 | 4 | 1 |
| Fifth, (" panhandle "), Goodwell | 160 | 5,000 | 7,000 | 4 | 1 |
| Total | 720 | $75,000 | $86,000 | 24 | 6 |

omingo, in the second judicial district. The Conners School, at Warner, in the second district, received classes in February, 1909. The schools for the remaining districts opened for work late in the fall of 1909, and attempted to carry on but two terms of the year's work.

The schools possess no dormitories, but they hope to receive appropriations for these at the session of the legislature sitting during the winter of 1910-11. Each school expects then to have a central plant worth about $30,000, exclusive of the land, except the school at Helena, which now has a building and equipment valued at $50,000, presented to the state.

The only courses offered are the agricultural and domestic economy courses. Each school maintains a three-year course in agriculture and home economics, and three years of "preparatory" work, that of the sixth, seventh, and eighth grades. No pupil is admitted to these grades who has similar privileges in his home district. Students over sixteen may take certain special courses. The older schools are also maintaining "short courses" of two weeks for farmers and their wives. These include instruction and demonstrations in domestic economy, canning, preserving, and cooking, for the women, and various agricultural subjects for the men.

The regular courses comprise for the three years approximately 150 periods of recitation work, and 75 hours of practical work, averaging for each week, 16½ recitation periods, and 8½ hours practice, though the latter is less in the third year than during either of the other two.

The school at Tishomingo enrolled for the year 1908-9, 9 in the second year, 24 in the third year, and 64 in the three years of the grade work (special course), total 97, averaging 17 years of age. These students were drawn from ten counties. The faculty numbered 6 teachers. The agricultural work was carried on by the principal and another agriculturist.

The school at Broken Arrow opened its doors in November, 1909, with an enrollment of 176, with 8 instructors and assistants.

CURRICULUM OF AGRICULTURAL AND DOMESTIC ECONOMY COURSES IN THE JUDICIAL DISTRICT SCHOOLS OF OKLAHOMA

[Figures in parentheses indicate hours of practical work per week. Letters *a*, *b* and *c*, indicate first, second, and third term.]

| Course in agriculture for boys | | Course in domestic economy | |
|---|---|---|---|
| Subjects | Hours per week | Subjects | Hours per week |
| | *First Year* | | |
| English | 5 | English | 5 |
| Arithmetic (*a*) | 5 | Arithmetic (*a*) | 5 |
| Algebra (*b* and *c*) | 4 | Algebra (*b* and *c*) | 4 |
| Physiology (*a* and *b*) | 4 | Physiology (*a* and *b*) | 4 |
| Civics (*c*) | 5 | Civics (*c*) | 5 |
| Stock judging (*a*) | (2) | Agricultural botany (*a*) | 3 (2) |
| Breeds of animals (*b* and *c*) | 3 (4) | Social culture (*b* and *c*) | 1 |
| Drawing | (2) | Drawing | (2) |
| Carpentry | (4) | Cooking | (4) |
| | | Sewing | (4) |
| | | Laundry (*b*) | (2) |
| | *Second Year* | | |
| English | 4 | English | 4 |
| Algebra | 5 | Algebra | 5 |
| History (*a* and *b*) | 4 | History (*a* and *b*) | 4 |
| Agricultural physics (*c*) | 4 (2) | Farm crops | 3 (2) |
| Vegetable gardening (*a*) | 3 (2) | Cooking | (4) |
| Soils and fertilizers (*b*) | 3 | Sewing | (4) |
| Farm crops (*c*) | 3 (2) | Farm crops (*c*) | 3 (2) |
| Drawing farm plans | (2) | Drawing | (2) |
| Agricultural practice (*a*) | (2) | Household art | 1 |
| Farm dairying (*b*) | (2) | Farm dairying (*b*) | (2) |
| Blacksmithing | (4) | | |
| | *Third Year* | | |
| English | 4 | English | 4 |
| Geometry | 5 | Geometry | 5 |
| Forestry (*a*) | 3 | Forestry (*a*) | 3 |
| Plant propagation (*a*) | (2) | Plant propagation (*a*) | (2) |
| Farm economics (*a*) | 3 | Invalid cooking (*a*) | (4) |
| Farm machinery (*a* and *b*) | 1 (2) | Domestic hygiene (*a*) | 2 |
| Road making (*a* and *b*) | 1 (2) | Home nursing (*a*) | 2 |
| Agricultural practice | (2) | Home management (*b*) | 1 |
| Fruit growing (*b*) | 3 | Domestic chemistry (*b*) | (4) |
| Farm dairying (*b*) | (4) | Millinery (*b*) | (4) |
| Farm accounts (*b*) | (4) | Farm dairying (*b*) | (4) |
| Entomology (*c*) | 3 | Home economy (*c*) | 1 |
| Feeding and management of farm animals (*b* and *c*) | 3 (2) | Entomology (*c*) | 3 |
| Diseases and care of farm animals (*c*) | 3 (2) | Floriculture (*c*) | 1 |
| | | Dressmaking (*c*) | (4) |

## *The Wisconsin and Michigan County Agricultural Schools*

The county schools of Wisconsin and Michigan stand in a class by themselves, not only in regard to amount of state aid given, but also in the character of the curriculum. They show very few traces of the literary or academic influence. They lack only the large amount of ground possessed by other schools and the farm work entailed to make them the most intensely vocational of all the publicly supported agricultural schools. Taking pupils who for the most part have finished the eighth grade, they plunge immediately into a variety of lines of agricultural theory. It is very seldom that we find the students of other schools taking up the various special phases of animal husbandry, such as stock judging and dairying, in the first year. The Guthrie County High School, of Iowa, is the only other important one here recalled that is a public high school. The

### CURRICULUM OF MARATHON COUNTY SCHOOL OF AGRICULTURE

[The letters in parentheses (*a*, *b*, and *c*) indicate fall term, winter term, and spring term, respectively.]

| *First Year* | | Periods |
|---|---|---|
| Plant husbandry: (a) Agricultural botany; (b) plant manipulation; field crops; (c) fruit growing; (d) gardening | | 3 (4) |
| Animal husbandry: (a) Breeds—dairy and beef cattle; (b) dairying, poultry; (c) breeds—horses, sheep, swine | | 3 (4) |
| Manual training: (a) Free-hand drawing; (b) joining; (c) cabinet making | | 0 (10) |
| English | | 5 |
| Geography (*a*) | 5 | |
| Arithmetic (*b*) | 5 | |
| Bookkeeping (*c*) | 5 | 5 |
| | | 16 (18) |
| *Second Year* | | |
| Plant husbandry: (a) Soil physics; (b) soil fertility; (c) special crops | | 3 (4) |
| Animal husbandry: (a) Feeding and breeding; (b) creamery; (c) stock farming | | 3 (4) |
| Manual training: (a) Forging; (b) rural architecture; (c) rural engineering | | 0 (10) |
| English literature | | 5 |
| Chemistry (*a*) | 5 | |
| History (*b* and *c*) | 5 | 5 |
| | | 16 (18) |

Class-room work, each year, 16 periods, or $10\frac{2}{3}$ hours.
Class-room work, two years, 32 periods, or $21\frac{1}{3}$ hours.
Laboratory, shop, and field practice, each year, 18 periods, or 12 hours.
Laboratory, shop, and field practice, two years, 36 periods, or 24 hours.

Marathon and Dunn County schools were the first established. Their courses of study may be taken as typical of all, the differences being merely in minor changes in the position of different branches. The curriculum of the Marathon County school is given in full, because it needs less changing than the others in order to make it comparable with other curricula given.

The following percentages show what proportion the class-room work in agriculture and related science is of the total class-room work, and what proportion all the agricultural work is of the total time required of the student in the class-room, laboratory, shop and field: Proportion of class-room work in each year, 37 per cent; proportion of all the time in each year, 71 per cent.

While practically the same subject matter is included in the Dunn County school's course of study, it is broken up into smaller units and scheduled fewer times a week. They are arranged with reference to four terms of eight weeks each. Below are given the percentages of the time occupied by the agricultural subjects, with a sample program for one of the eight terms.

TABLE 48

TIME GIVEN TO AGRICULTURE IN PROPORTION TO OTHER SCHOOL WORK

| | | First term | Second term | Third term | Fourth term |
|---|---|---|---|---|---|
| | Per cent | Per cent | Per cent | Per cent | Per cent |
| First year: | | | | | |
| Class-room work | 47 | 33 | 33 | 70 | 50 |
| All time | 75 | 72 | 72 | 85 | 75 |
| Second year: | | | | | |
| Class-room work | 84 | 86 | 86 | 76 | 88 |
| All time | 90 | 93 | 93 | 76 | 90 |

PROGRAM OF THE DUNN COUNTY SCHOOL FOR THE THIRD TERM OF THE SECOND YEAR

| | Periods |
|---|---|
| Literature | 2 |
| Library reading | 2 |
| Hogs | 3 |
| Seeds and germination, green house | 1 |
| Shrubs, trees, and planting | 3 |
| Birds and insects | 2 |
| Emergencies | 4 |
| Creamery practice | 2 |

## Faculties of the Special Agricultural Schools

There seems to be no uniformity of practice in choosing principals for the special agricultural schools. The southern schools do not incline to elect agriculturists to these positions. In only three of the nine Alabama schools is the agriculturist also the principal. One principal reported that he taught four other classes a day in addition to those in agriculture. Six of the 11 principals of the Georgia schools are not agriculturists, but teach the non-industrial branches. Of the remaining 5, only 3 are plainly listed as teachers of agriculture, 1 is "principal and superintendent" (with 5 assistants), and 1 is "principal, assistant to all the departments, and general supervisor and director." The county schools of Michigan and Wisconsin, with one exception, have agricultural college graduates for principals. One of the Oklahoma schools has an agriculturist for principal.

The faculty roll of the Crookston School of Agriculture shows a superintendent and instructors in academic branches, mechanics, home economics, poultry, dairying, and music.

The California Polytechnic School, with a faculty of 16, besides the farm foreman and dairyman, has a specialist in English as principal and a mechanical engineer as vice-principal. The former principal was an agriculturist. About half of the students take the mechanics course, while the other half is divided almost equally between agriculture and home economics.

Incomplete returns from the so-called "district agricultural schools" in Virginia point to a strong tendency to appoint agriculturists as heads of agricultural departments rather than as principals of the schools. This may properly enough be in recognition of the fact that these schools serve as college preparatory schools and as normal training schools for their respective localities, as well as agricultural schools.

Thirteen agriculturists, who are principals of as many special schools, receive from $1,000 to $2,400, averaging $1,723. A somewhat smaller number, serving as assistants or instructors, receive from $583 to $1,800, averaging about $1,100. The two extremes, in this case, were found in the same school. In only two instances, was this average for the instructor exceeded by the salaries of agriculturists teaching in public or general high schools, in one case by the salary of a man engaged for a limited period to organize the school, and in the other instance, by that of an instructor in a county high school of the Northwest.

The number of men and women teaching in these special schools is shown in Table 49. In this table is also shown the number of persons in each state who are teaching the industrial subjects. Especially noticeable is the disparity between the Alabama schools with but little more than 20 per cent of their teachers so engaged, and the Wisconsin schools with over 85 per cent.

TABLE 49

FACULTIES OF THE SPECIAL AGRICULTURAL SCHOOLS FOR COUNTIES AND CONGRESSIONAL DISTRICTS

| State | Number of schools | Teachers employed | | | Teachers of industrial subjects | |
|---|---|---|---|---|---|---|
| | | Men | Women | Total | Number | Per cent |
| Alabama | 9 | 27 | 15 | 42 | 9 | 21 |
| Georgia | 11 | 35 | *a*18 | 53 | 31 | 59 |
| Michigan | 1 | 2 | 2 | 4 | 3 | 75 |
| Oklahoma | 6 | 24 | 6 | 30 | 22 | 73 |
| Wisconsin | *b*4 | 9 | 6 | 15 | 13 | 87 |
| Total | 31 | 97 | 47 | 144 | 78 | 54 |

*a* One of these gives possibly less than half-time to teaching.
*b* One school not reporting.

## CHAPTER VII

# PROBLEMS OF AGRICULTURAL INSTRUCTION IN THE SECONDARY SCHOOL

### Agriculture and the Existing School System

To insure an efficient general system of agricultural education of the secondary type, it is necessary to adopt a policy that shall be adequate and far-reaching, and which shall remedy the defects of our schools as they exist to-day without diminishing their usefulness. If we may judge by the discussion now going on, the determination of such a policy is a matter of considerable difficulty.

We have at the outset, in order of time, the proposition to make the teaching of agriculture in the elementary schools compulsory. This has not proved as satisfactory as its proponents had hoped, and would probably be no more so in the existing high schools.

Instruction in manual training has been given a decided impetus in some states by state appropriations, especially where made dependent upon the amount of the local expenditure. It seems reasonable to suppose that some such stimulus would greatly promote industrial work of the pronounced rural type. While it has not done so to any considerable degree in Maine,[1] it has been very successful in Minnesota and Virginia. The most notable application of this principle of state encouragement has been the aid given to the special agricultural schools, where different localities in a congressional district have competed with each other in offering land and money to furnish the plant of a school for the district, and where counties have continued to share the running expense of county schools in order to insure their continuance. The requirement for a stand-

[1] See Chap. II, p. 21.

ing in agriculture on the elementary teacher's certificate has roused many high schools to the necessity of making as great an effort to help their graduates to "pass" in this subject as they make in the other common branches by the so-called "reviews." In certain southern states this has been, in the ordinary high school, merely book work of the most formal type. In such states as Nebraska and Michigan, where 150 or more high schools have normal training classes, the preparation in agriculture is often of a grade that will compare favorably with the preparation to teach the other subjects, and has in it a large observational and experimental element. In New York state the same forces are beginning to exert a strong influence, though probably not yet resulting in such efficient work, on the average, in the forty or more training classes that have undertaken to give some work in agriculture.

Aside from the question of the efficiency of agricultural teaching in particular, there are probably 200 or 300 small high schools with not more than two teachers each, mostly in Missouri, Nebraska, and Ohio, in which agriculture is taught as well as any of the other sciences in the same schools. Consequently any criticism of the agricultural work in these schools must lie against the school as a whole and not against the subject. Furthermore, in Nebraska there is a large number of first-class, well-equipped high schools, in which agriculture is probably taught as well as the other sciences that pass muster with the university inspectors, and better than it is in many normal schools. The work of these schools and of other more widely scattered cases in New York, Indiana, Michigan, Minnesota, North Dakota, Iowa, Kansas, Utah, and California, demonstrates conclusively that agriculture can successfully take its place as a year or as a half-year study with the other branches of the high-school curriculum.

The work of township schools, like those of Petersham, Mass., Waterford, Pa., North Adams, Mich., and the John Swaney School in Putnam County, Ill., and such county public high schools as those at Calvert, Md., Panora, Iowa, and Dillon, Mont., has shown conclusively that agriculture can be made the core of a four-year high-school course, on a par with the classical or the English-scientific course, commanding more respect from the authorities than the average commercial course, from both the vocational and the cultural standpoint. Whether the usual public high schools can do such work as well as special schools is a matter for the future to decide.

### *The Effect of Establishing Special Schools*

The exponents of the special agricultural school urge that this is an age of differentiation, of specialization, and that an agency devoted to one purpose can fulfil that purpose better than one that scatters its energies. The opponents urge, on the other hand, that the existing high-school system is sufficiently elastic to do this work as effectively as it has accomplished other new lines of work delegated to it, and that if the same state or national aid is granted to these schools as is proposed for the special schools not yet created, they will accomplish the same results, and without duplication of plant, of administrative machinery, or of teaching force in such lines as will inevitably be taught in both kinds of schools,—such studies, for example, as English, history, civics, mathematics, and possibly modern languages. The opponents of the special school further urge that the creation of such schools exclusively for rural pupils will take away much needed support from schools now doing good work, but which depend for their success upon the combined support of the village and the surrounding township. They point to the successive annual reports of the state departments of education to show how rural high schools that were in the third grade are now in the second grade, and how others in the second grade are now classed as first-grade high schools, and claim that this advancement would have been impossible had there been special agricultural schools of secondary rank nearby to draw off this important clientage. Reference to the figures given on the first page of Chapter II will help one to see that there may be some force in this contention. It should be remembered that the total there given includes the data from many city schools with a low percentage of attendance from rural districts.

Many readers will recall that this argument has been advanced against the normal school's giving academic instruction of secondary grade. Under certain circumstances this would seem to be a valid objection to the special secondary school. The strength of the objection will depend somewhat on the following two conditions: Whether pupils are to be able to enter both schools with the same preparation, or possibly able to enter the special school with less previous schooling, and whether the two types of schools enter into active competition for students in the same territory by reason of the special school offering as wide a range of studies as the high school; for a few do so even now. It has been pointed out by Dr. Thorndike[2] and others that the small high school could do much better work by concentrating its energies on two years of good work instead of spreading them over four years of work very indifferently done. With the same thought in mind a very sane proposal has been outlined in detail by Assistant Secretary Hays,[3] of the U. S. Department of Agriculture. He urges that when the ungraded district schools shall have consolidated they shall offer two years' work above the eighth grade, that shall be of a general nature, and include agriculture in its more elementary phases. Thus these schools will provide communities with two years of high-school work that previously have had none, and will not exhaust their resources by trying to maintain too pretentious a course. This has already been done in many rural districts. The scheme further provides that the next two years' work shall be given in a central school for a large district, say, of ten counties, amply equipped to do strong work along specialized lines, namely, agriculture, manual training, and home economics. While no definite suggestions have appeared as to means of preventing a competition disastrous to these two-year high schools, such central agricultural schools could control it without friction by refusing to admit pupils into corresponding grades who come from townships provided with such ten-grade schools, except under exceptional circumstances. This would

[2] A Neglected Aspect of the American High School, *Ed. Rev.*, March, 1907.

[3] Education for Country Life, Office of Experiment Stations Circular 84.

follow the precedent set by the special schools of Oklahoma. But so long as state universities and state departments of education place such manifestly weak four-year schools on their accredited lists as they occasionally do, even now, so long must we expect local pride to insist on the three- or four-year high school, even though it has but one teacher, with the consequent outcry against the special agricultural school as the enemy of the present public-school system.

The general high schools maintained by counties and a few of the wealthier townships present a somewhat different problem. There is no doubt that some of these are doing very strong work, a few of them already having experimental plats of several acres. Personal observation has justified the conclusion that along agricultural lines instruction just as pedagogical, equipment just as good, and instructors just as capable, are sometimes found in general public high schools, as in some agricultural schools existing in the same towns with high schools not teaching agriculture; and that too in spite of the fact that the non-specialized high schools teaching agriculture received no state subsidy. But we find a parallel to this seeming difficulty of local high schools duplicating the work of a large district agricultural school in the existence of city normal training classes, and county normal training schools working in the territory tributory to state normal schools with no jealousy or waste of effort, as each ministers to a somewhat different need, and both do not succeed in supplying the demand for trained teachers. New York with 13 state normal schools, has about 70 local training classes; Nebraska's 2 state normal schools are supplemented by more than 100 high-school training classes; Michigan with 4 state normals has over 40 training classes; while 24 counties of Wisconsin find that her 7 state normals are unable to meet their demands. Whether the establishment of special schools would discourage the introduction of agriculture into the regular high schools of the district is a matter upon which at present we can not argue from known facts.

The reader will note that all of the considerations mentioned above are purely of an administrative nature. Important arguments of another sort have been urged both for and against the special school; on the one hand that the concentration of interests in one place would make it more efficient, and that its location should be in a rural environment (as many high schools located at county seats are not) both for the sake of better facilities and of an agricultural "atmosphere." On the other hand, we hear that such schools will be undemocratic, that they will set up (or down) class distinctions, placing a hampering barrier about those who are trained in them, and that their students will be deprived of the culture they would gain in the present high schools. But it is open to question whether many of the cultural studies should be presented to students looking forward to an agricultural career in the same manner as they usually are now. Mathematics and the sciences are now taught, and are treated in the texts, if we may believe the teachers and authors, very largely as cultural subjects. Indeed these persons often seem jealous of the standing of history, literature, and the languages as being cultural subjects. A pertinent suggestion has been advanced by State Commissioner Snedden of Massachusetts that while segregation of agricultural instruction may not be advisable as a permanent policy, the special schools may be much better able to work out a suitable treatment of the entire range of cultural studies by themselves than may the general high school hampered by its traditional standards. Such a development might call for the introduction of material on the history of industrial interests, for the study of political economy and the physical sciences with special reference to agriculture, and for a reconstruction of secondary-school mathematics. As yet only a few signs are visible that this is really happening in the present technical schools of agriculture. It takes a man of broader learning to reconstruct the humanities along these lines than the special schools are yet attracting. The agricultural high schools occasionally list "agricultural" botany, chemistry, or physics. It seems necessary at present for most of them to use a standard text supplemented by some technical reference work. In two or three cases the instructor is making his own text-book. Only one or two of the many catalogues at hand show anything but the regulation courses in algebra and plane and solid geometry. The exceptions are efforts to introduce work in farm accounts, farm surveying, laying out of fields, and the determination of slopes for the purpose of fixing drainage lines. Just how much of the algebra and geometry can be "agriculturized" is not apparent at first sight. Unless we propose to throw out all that is inapplicable, that is, the bulk of it, the present mathematics must evidently hold its place for a supposed disciplinary purpose.

If we grant the desirability of maintaining separate agricultural schools in order that they may develop untrammelled their own body of cultural material, it is still possible, if we keep the goal plainly in view, to bring about an ultimate articulation between them and the rural high schools along the lines suggested by Assistant Secretary Hays. The only procedure necessary would be for the special school to "raise the standard," that is, to lop off its lower grades as soon as the local schools seem to be able to offer all the agriculture that the younger pupils of the high school may advantageously pursue. That this can be done is shown by the action of the Alabama schools in dropping the work they formerly carried on below the seventh grade. As General Hancock remarked about the tariff, agricultural education is largely a "local issue." Both the subject matter and the administration should be conditioned by the environment. One is told in Iowa that neither the boys nor the community could be interested in poultry—they were "too busy getting rich to bother with chickens." Nor would they have been much interested in the Michigan boys' work with the polariscope, except as a toy. Iowa was not one of the states, however, that reported "animal feeding" as one of the difficult topics of instruction. So the form of organization of the facilities for teaching agriculture might well be modified in different sections of the country to meet existing local needs, and to articulate with the existing school system.

The principal difference of opinion among educators regarding the early drafts of the so called Davis bill has centered around this point. In its present form[4] it proposes "to co-operate with the states in encouraging instruction in agriculture, the trades

[4] H. R. Bill 20374, 61st Congress, 2nd Session, introduced by C. R. Davis, of Minnesota, February 8, 1910.

and industries, and home economics in secondary schools; . . . " by appropriating pro rata $4,000,000 to aid state district agricultural schools, $1,000,000 for experiment stations attached to them, $5,000,000 "for the maintenance of instruction in trades and industries, and home economics and agriculture, in public schools of secondary grade," and $1,000,000 for similar instruction in state normal schools. The agricultural schools are to be not less than one for every fifteen counties nor more than one for every five counties or fraction of five counties. The bill as at present drawn gives great freedom to the proper state authorities to designate any suitable high school to receive the aid. An objection to the first draft was that it merely stipulated "cities," without fixing the lower limit of the population of municipalities free to avail themselves of the grant. An objection to the second bill introduced was that the federal aid would be open only to high schools of "incorporated villages, towns (not townships), and cities containing at least two thousand inhabitants," and even then could "be used only . . for the distinctive studies in mechanic arts and home economics," that is, for "instruction in the non-agricultural industries and in home making in the city secondary schools." How this would have discriminated against many of the high schools now attempting such instruction, is indicated by the fact that about one-half of the schools in which agriculture was taught when Mr. Davis's bill was introduced would not be eligible to receive any federal aid. It should not be claimed, however, that all of these did work deserving recognition, or even that all would do creditable work if given a grant.

As already hinted, the contemplated special schools could partially supply the deficiency of small but well supported high schools in the South and supplement them in the North and West.

One of the arguments raised against the special school has been the disadvantage of sending the youth from home, both on account of their removal from the influence of the farm life while away at school, and on account of the evil influences and distractions of the city. In reference to this point, more or less complete views were obtained from the principals of twenty-two special schools of all types in response to the following questions: (1) Do you seem to see any undesirable effects caused by town attractions on those pupils living away from home? (2) Do you think it would be better if all could (a) live at home and attend as "day scholars," or (b) board in town, or (c) be housed in school dormitories?

Fifteen respondents expressly favored dormitories and two others apparently do, as their schools are erecting such buildings; two found the village accommodations very satisfactory; and one favored students living at home. Only two seemed to notice bad effects from village associations. Ten failed to see any such effects, often remarking on the essentially rural character of the location of the school. Five others stated that their schools were in the country, so that the question did not apply. One believed the school should be near a city for the sake of cultural advantages; his school is one of those now providing a dormitory.

The summaries given in the following table are pertinent to the above remarks.

TABLE 50

DATA ON STUDENTS IN AGRICULTURAL HIGH SCHOOLS

| States | Schools | Dormitory | Boarding | From farm homes | Enrollment |
|---|---|---|---|---|---|
| | | | Per cent | Per cent | |
| Alabama | 7 district | no | 50 | 75 | *a*1,008 |
| Georgia | 11 district | yes | *b*77 | *c*87 | 1,001 |
| Michigan | 1 county | no | 20 | 71 | 48 |
| Minnesota | 1 state | yes | 92 | 95 | 63 |
| Virginia | 1 district | no | 50 | 50 | 41 |
| Wisconsin | 3 county | no | 60 | 78 | 191 |

*a* Two not reporting.
*b* Based on the enrollment of 750 in the eleven schools for the fall term 1909.
*c* 763 of 874 enrolled in eight schools during the year 1908–9.

## THE ATTITUDE OF COLLEGES AND UNIVERSITIES

While public attention seems to be focused on the relation of agricultural education of secondary grade to the present system of secondary education, its relation to the higher institutions is of no small importance. Its dignity and standing in the community will depend to a greater or less degree upon the recognition given it by the colleges and universities. Cases are known where schools have abandoned the work because it could not be presented for entrance at the state university, although meeting with favor among patrons and pupils.

The separate agricultural colleges, and the agricultural departments of state universities as well, have not, as a rule, set up as stringent entrance requirements as have the private literary colleges and the "liberal arts" colleges of the state universities. The agricultural colleges have been filled with a desire to use their plants, the only facilities for agricultural instruction which the states have had until recently, to their utmost capacity for the good of their constituencies. While they have taken students with less scholastic attainments than have the literary colleges, the agricultural colleges still have had to take them from the same public school system. Few if any conditions have been prescribed that would not be required for entrance into the literary colleges. The agricultural colleges have gladly taken any farmer boy with a classical course, and with no high-school science, for the village schools probably teach the classics less badly than they do or would teach science. Nearly every state university is liberal about accepting various combinations of well-taught high-school sciences, with certain minimum requirements in history, language, and mathematics.

Agriculture as a high-school subject is comparatively new. The agricultural colleges could not well avoid accepting it as an entrance subject and have, with a few exceptions, gladly done so. The departments of arts, letters, and science, of the state universities, have, however, looked upon it with suspicion, while private colleges have, almost without exception, refused to have anything to do with it. Within the last two years the college attitude toward the subject has grown much more favorable. Up to this period probably not one of the larger universities could be found willing to accept it for entrance to any but the agricultural college. So rapidly is this change in feeling going on that one set of responses regarding the official recognition given agriculture is scarcely all in before the information is out of date and unreliable.[5] In most cases such information has been received from the accrediting officer of the institution, though sometimes from officers of instruction. The following statement shows the recent attitude of the leading universities of the upper Mississippi Valley, and of a few others outside this territory. It will be noted that several institutions have not had to face the problem of passing on this question, but are inclined to allow the subject an opportunity to justify itself.

*Ohio.* Any college of Ohio State University conferring degrees will accept year and half-year courses based on Bailey's or Jackson and Dougherty's texts.

Miami University will accept work based on the latter text.

*Indiana.* The subject has never been presented for entrance to the University of Indiana, but would probably be accepted from a "commissioned high school" teaching it as one of the four regular studies constituting a year of high school work.

Purdue University. "Purdue has no specific arrangement by which high school agriculture is accepted as entrance subject, yet it is tacitly understood that the botany offered for entrance may be agricultural, in fact from some of the high schools it is largely so." (This is the state agricultural and mechanical college of Indiana.)

*Illinois.* The various degree-conferring departments of the University of Illinois will accept year and half-year courses.

The University of Chicago would probably accept "scientific" agriculture, in which the underlying principles are studied by laboratory methods in and out of doors.

*Michigan.* The catalogue of the University of Michigan does not include agriculture among the subjects that may be presented for entrance credit. The dean of the department of literature, science, and the arts thinks that credit should not be allowed for it in this department "in that we teach no agricul-

[5] Consult the extensive list compiled by the members of the committee appointed by the department of rural and agricultural education of the National Education Association to investigate the question of college entrance credit in high-school agriculture, and reported at the Boston meeting, July 7, 1910.

ture in this department," and he is "unable to see how a course in agriculture would enable a pupil to take up university work advantageously."

*Wisconsin.* One-half year's work may be presented for entrance to the University of Wisconsin under the category of "optional work." It may also be presented together with botany for a year's credit.

*Minnesota.* As yet this work may be presented only in the college of agriculture. The university strongly recommends that high schools place large emphasis upon the agricultural application of all the sciences taught in the high schools.

*Iowa.* The State University of Iowa does not accept agriculture because "the content and character of the courses in the very few instances where they are given is so uncertain . . ." Exception would doubtless be made to the general rule "in the case of a given school in which a course in agriculture had become well established under proper conditions including an especially prepared teacher, who is a master of the subject not only from the content side but also from the pedagogical side." The inspector sees no reason why such well organized work "should not be credited as substantial high-school work." The principal of the Guthrie County High School reports that the university inspector regarded the agricultural work in the school as being of the same grade as the other science work.

*Missouri.* The agricultural college and the teachers college are the only departments of the University of Missouri that accept agriculture at present. Action by the faculty of the college of arts and sciences seems to wait upon the formulation by the agricultural college faculty of a unit of agriculture, which they had not done at the time of the latest information received, July, 1909. The announcement for the summer session for that year mentions, however, that certain work offered in agriculture will be counted toward the B. S. degree in the teachers college, and also that any three of these same courses will be accepted by the college of agriculture as one unit for entrance.

*Kansas.* In the catalogue of the University of Kansas, agriculture is listed as a half-unit course in the group of industrial subjects, from which one unit may be offered for entrance to any course in the college of arts and science leading to the bachelor's degree. But the high-school agricultural course must first be approved by the university high-school visitor.

*Nebraska.* In the University of Nebraska, a half-year's credit will be allowed for agriculture offered for entrance into any course leading to the A. B. or B. S. degrees or to the degrees in pharmacy.

*California.* The University of California will accept a half-year's work in dairying if presented with chemistry, or a half-year's work in horticulture if presented with botany, providing the agricultural subjects follow their accompanying sciences during the third or fourth years of the high-school course. In order to encourage introductory science in the first year of the high school, the university will credit a year's course, the outline of which recommends, among other topics, " such elementary scientific principles as are involved in gardening, including a study of soils, and elementary physiography, the weather, simple machinery, including the steam engine, . . ."

*New York.* Cornell University has left the question in abeyance until it " should become a practical one through an application for admission to our college by some candidate proposing to offer the subject of agriculture. On the existing state of the facts such a candidate [in arts and science] would unquestionably be credited with that subject. Whether we should then continue to accept it or should take action similar to that already taken by the law and engineering colleges [refusing to accept it], I am unable to say."

*Alabama.* " The entrance requirements of the University of Alabama are now being modified so as to include agriculture. Next fall [1909] one unit may be presented for admission."

*Georgia.* One or two years of work in agriculture may be presented for entrance to any department on a par with other sciences.

*Tennessee.* The 1908 catalogue of the University of Tennessee gives one " point " (Carnegie Foundation value) for agriculture. The requirements include an elementary text, such as

Burkett, Stevens, and Hill, with practical demonstrations and experiments in the school garden, or on the farm. The respondent also adds "practical experience on the farm is also counted, and if of two or more years may be sufficient without the text book."

*Virginia.* Agriculture is not accepted for entrance by the University of Virginia, although the subject is taught in its summer school for high-school teachers.

The relation of some of the above facts to local conditions merits some attention.

The attitude of the universities, with one exception, is as advanced as the state of agricultural instruction in the high schools of the respective states, and in many cases is more so.

Scarcely any candidates seem to have offered agriculture in the literary department of universities which would accept it; and the other colleges have not been called upon even to pass on the question. Possibly this is because the young men who could offer agriculture go to the agricultural colleges if to any.

The number of high schools on the accredited list teaching agriculture has been very small in most states. Until the past year there have not been more than two or three each in Illinois, Indiana, Iowa, Michigan, Minnesota, New York, and Tennessee, and scarcely more than half a dozen in Kansas and Wisconsin. (The University of Kansas but lately reported favorable to accepting agriculture.) A majority of the high schools of Indiana and Missouri teaching agriculture are the two-year and three-year high schools. Most of those in Ohio, even though four-year high schools, are one- and two-teacher schools. Nebraska has for some years made the best showing both in the number of schools teaching agriculture, and in the ranking of the schools doing so. The status has been changed materially within the past year by the introduction of agricultural departments in ten Minnesota high schools, and of a four-year course in agriculture in eleven Michigan schools. Several of the eighty or more New York high schools with agriculture in their training classes are also teaching it as a part of the regular high-school course. A number of Illinois high schools have recently taken teachers from the agricultural college.

The California proposition is suggestive in many ways and is a distinct contribution to the problem of the introductory science in the first year. The Tennessee plan is interesting, but probably represents only a transitional phase.

The action of one faculty was much more advanced than the ideas of the head of the department of botany, the line of science most closely connected with high-school agriculture. Several professors of botany expressed their willingness to accept high-school botany with a decidedly agricultural content up to a third or half of the total work done, provided the combined courses extended through an entire year, which seems a reasonable position for them to take.

Below are found the names of subjects that are required to receive one unit or one-half unit credit in elementary agriculture on college entrance requirements in Ohio State University. " While in our printed catalogue, our college entrance units do not have such a lengthy wording as this, yet it is sufficiently comprehensive to include the subjects here named:"

Elementary Agriculture Accepted as Entrance Credit by Ohio State University

One unit:

- One year given to "Agriculture Through the Laboratory and School Garden," by Jackson and Dougherty; or " First Principles of Agriculture," by Bailey.
- Special attention should be given to the plant and its relation to the soil—
  - The preparation of the seed-bed and germination.
  - Soil moisture and temperature.
  - Drainage and conservation of soil moisture.
  - Plant foods.
  - Plant propagation.
  - Plant improvement.
  - Selection of fruits, vegetables, and cereals best adapted to climate, soils, home use, and markets.
  - These subjects should be accompanied by recorded experiments and observations.
- Special attention should be given to the relation of animal forms to plants—
  - The beneficial effects of insects.
  - Insect pests and insecticides.
  - Types of farm animals and their characteristics.
  - Care and characteristics of animal products.
  - Common scale forms, insects, and types of farm animals should be identified.

One-half unit:

- One-half year's work given principally to the plant and its relation to the soil.
- Recorded experiments and observations.

## Difficulties of the Curriculum

### *Relation of Agriculture as a Branch of the High School Curriculum to the Sciences Already Present*

By far the larger part of elementary agriculture, as judged by the amount of space given in text-books and syllabi prepared by the school authorities, is made up of the plant phases of the subject. Crops and soils, forage crops and feeds, the garden and the orchard, these are the things that are mostly considered. Undoubtedly most of these lend themselves more easily to field observation and laboratory study than do such topics as breeds of animals, farm buildings, or good roads. The scientific principles underlying these dominant topics are largely the underlying botanical principles of plant structure, plant nutrition, variation of seedlings, and inheritance of characteristics. The line of pure science taught in the colleges that comes into closest relation with public school agriculture is botany, probably much more so than its close competitor, chemistry. The attitude of the heads of the departments of botany in the leading universities of the rich agricultural states, is not only of interest; it is of importance. As might be surmised, the most cordial sympathy is found mostly in those state universities with agricultural departments, the most conservative note in the universities of states maintaining separate agricultural colleges. The following questions were asked of the professors of botany in a number of the universities:

"Would botany be acceptable (for entrance) that laid a great deal of stress on such agricultural topics as corn judging, seed selection, production of new varieties by selection and cross breeding, pruning, and grafting, plant diseases and their treatment, experimentation with fertilizers and soil treatment to determine effects on plant growth, weeds, etc.? Reference is made, of course, to such work as deals with these topics in the laboratory, field and orchard, not to book work. Do you consider such work as mentioned desirable in the general botany course in a village school?"

The answers are all more or less tinged by various conceptions of the term "agriculture" considered as a high-school subject.

Two writers, viewing it as a purely vocational subject whose work must necessarily be given by empirical means, intended to teach the art, express views strongly opposed to it. One admits, however, that "agriculture following botany, zoology, and chemistry, might be well and good," but thinks it "folly to attempt to teach children corn judging, etc."

A third writer, believing that "we shall be confronted with the question in the near future," thinks that "there are certain topics pertaining to agricultural science which might very well receive recognition on the part of the universities, but there are also very many which are so far away from the ordinary conception of educational work that it seems to me that their evaluation would be a matter of serious difficulty."

It may be interesting to compare the above views representing the leading state universities without agricultural colleges with those expressed by an influential member of the faculty of the University of Chicago, the most important privately endowed university of the Central States, and one that resembles the others mentioned in not maintaining a college of agriculture. Professor John M. Coulter distinguishes between agriculture as a science and as an art, and expresses the opinion that "the former deserves to be accepted for entrance, the latter could not be.[5] The contents of botany, as you list them, are all right, provided the work is based on the reasons for things, and is not merely empirical. It is the attitude of mind toward the work rather than the work itself that determines its worth for college entrance. The topics mentioned are very desirable in a country high school, but not to the exclusion of other topics . . ."

Heads of the department of botany in three state universities having agricultural colleges agree that from one-fourth to one-half of a full year's course in botany might profitably be spent on many of the topics enumerated, provided that botany in the

[5] As previously noted, the University of Tennessee will accept certain agricultural knowledge based upon experience instead of on the texts usually followed.

strict sense be scientifically taught before such topics be taken up. One thinks they would be a good substitute for much of the plant analysis work where the course runs throughout the year. (The tendency in his state, is strongly against allowing more than half a year to botany.) Another would "not be willing to accept corn judging, seed testing, . . . etc., *as botany*" although "all this would be very good if properly treated." He does not think that plant diseases and their treatment, and experimentation with fertilizers can be treated "in any other than an empirical manner in the high schools."

The attitude of two other state universities of the upper Mississippi Valley, as ascertained through their high school inspectors, was to the effect that "botanical work given with an emphasis on the plant life common to farm life would be accepted provided it were done as scientifically as any other type of botanical work."

It will be seen that the botanists are generally agreed on these propositions: (1) The rural applications of botany are more or less commendable after the pure science has been presented; (2) They should form a minor part of the course. The views of several, and the policy of their institutions are to favor the segregation of agricultural work instead of its inclusion with botany. A note of dissent was uttered as to whether the topics suggested could be taught scientifically, i. e., according to the usual procedure of experimentation. Two of the writers objected to the idea that a high school should give any vocational training, even when resting on a foundation of science, going so far as to suggest that instruction in agriculture was as much out of place as instruction in forestry or pharmacy.

It will be noted that Professor Coulter is possibly the only one who would be willing to sanction the idea that the agricultural applications should form an integral and essential part of each topic of the course in botany, where such an application is possible. Most of the others express opinions that would not be consistent with the idea that the home environment of high-school students in rural communities should furnish the impelling motive for the study, a motive that should be consciously recognized by the student as furnishing the reason for botany's being in the course at all. Of course this is rather opposed to the so-called "cultural" and "disciplinary" views of education, views that the scientists seem strongly opposed to allowing the classicists to entertain all to themselves.

It is a fact of great significance that a number of high schools that have come directly under personal notice in this study have thrown out the subject of botany altogether (so far as one could tell by their statements) and have substituted agriculture, on the ground that neither the students nor patrons saw any sense in teaching botany but did recognize the value of scientific agriculture, especially when they saw results. They discarded a book using the name botany only to substitute for it an "agriculture" that treated of the structure of the flower, the method of pollination, the effect of cross-pollination of different strains of corn, with plans for field work to be done by the pupils at home on "corn breeding." There can be no doubt that this spreading tendency noticeable in small high schools is a protest against the formalism into which botany, in company with physics, has fallen. The botanists have only themselves to blame for the widespread substitution of a body of knowledge, poorly digested as yet, for a wholesome and scholarly kind of science work, because the influence of these leaders (as they should be) has been so largely for the "pure science," botany and so little for the kind that touches the life of the pupil and the interests of the community.

The exponents of the "new physics," the "new botany," etc., maintain that these subjects should come closer to the outside interests of the pupils and patrons of the school, and that consequently the sciences now in the curriculum can and should be so taught as to satisfy all demands for agricultural instruction that may legitimately be made upon the average public high school. But most of the good teachers of science are in the city schools. They find that it keeps them rather busy to bring the steel industry, baking powder manufacturing, and landscape gardening into their schools. The less efficient science teachers are in the one-, two-, and three-teacher high schools, and their texts say nothing about hydraulic rams, application of force to different parts of a plow beam clevis, the composition of fertilizers, or the purity of paints. They do not seem able to stir up much enthusiasm about sporophyte versus gametophyte generations, nor does the structure of four-o-clock seeds seem to create excitement. And the same texts do not mention the difference between kernels of corn that sprout and those that do not. Perhaps the botanist at the university does not care about it either.

It is clear that if the regular sciences are to meet the demands made upon them, the science teachers of small high schools must have more help than the present texts give them or their university courses furnish. When texts appear, as they are doing, that enable the teachers to make some use of their knowledge of the sciences and of their practical farm experience, when they have had any, it is only following the line of least resistance to follow the guidance of such a text or manual of laboratory and field exercises, and not attempt to make over the science texts already in their hands. Text-books are made to sell, and the small high schools can never be large users of texts in the special sciences. It may be too much to expect, as yet, that private enterprise will furnish rural editions of chemistry, physics, or zoology. The later agricultural texts partly perform this function for botany, with minor amounts of other matters thrown in. Even though technical agriculture be introduced into the upper years of the high school instead of the "elementary agriculture" now the vogue in the lower grades, the various sciences should so lend themselves to agricultural treatment as to free the technical subjects of the third or fourth years from enough pure science topics to permit an earnest study of the serious problem at hand.

The movement that has started in various parts of the country in very tentative fashion to work out an "elementary science" course[6] has taken on a unique form in California, where the state university plans a first-year high-school course, for which it gives a year's entrance credit, which shall include a variety of agricultural and general topics of nature and science. Over one-half or two-thirds of the work is of direct interest

[6] C. E. Peet, What Shall the First-year High School Science Be? Proc. of the N. E. A., 1909, p. 809.

to rural communities, whether pertaining to plants, engineering, soil formation, or meteorology. No doubt it is the facility with which agriculture has lent itself to this elementary science idea that has made it so popular as a first-year subject, where it manifestly could not become very technical. Given in the first year it must of necessity be largely cultural in its effect. It furnishes the opportunity for an introduction to the simplest chemical phenomena, combustion, solution, neutralization of acids in soil, the nature of nitrogen, of the meaning of the term protein, salts, and numerous other terms that any farmer must be slightly acquainted with in order to read intelligently his farm journal or the government bulletins, and which he may never hear of as a boy if they are left buried in a formal third- or fourth-year study.

The combinations of studies listed in Chapter IV are suggestive of the trend toward the use of a body of fairly simple facts and phenomena to fulfil such a function as just mentioned. Where once physical geography was expected to do this, we now find it combined with a half-year of agriculture. Unfortunately there is little evidence that the physical geography is modified at all by the relationship. At other times we find the popular combination of botany (flower study) in fall and spring, with agriculture (experimental work) in the winter. Undoubtedly the most efficient arrangement will result from a breaking up of the rather divergent lines of agriculture studies, so that plant work, such as the study of field, orchard, and garden crops, may be intimately taught with the principles of botany, when feeds and fertilizers will be integral parts of chemistry, when the cream separator will be the starting point of centrifugal action instead of the end, if indeed, the machine is not ignored altogether. The university botanist and chemist might not recognize their children in such a grouping, but the children would no doubt be lustier and grow to be more useful by the arrangement. This grouping will possibly grow more frequent as the "introductory science" comes more and more to be presented in the seventh and eighth grades of consolidated or village schools. Likewise the arrangement, not infrequently used, of this general course in agriculture in the fourth year, may also prove to be a temporary expedient, lasting only until the different sciences and their immediately related agricultural topics are welded together and rescued from pedagogical chaos. At such a time we may see the general high schools presenting, with fairly competent teachers, courses in science in appropriate years, so strongly "agriculturized" that they might bear indiscriminately the names of the present sciences, or the terms, agronomy, horticulture, farm mechanics, etc. With this state of teaching the present first-year agriculture would be largely relegated to the grades, as agriculture, nature-study, or elementary science according to the taste of the writer of the course of study. The more serious technical courses, requiring expensive equipment, large observational and experimental facilities where real plant and animal breeding and crop rotations may be studied under observation, may be taken in special schools not so far from the farmer, in time or place, but that he may see the results and profit by them himself as well as send his more ambitious and reliable boy thither for still more direct instruction. When we bear in mind that nearly one-half of our agricultural colleges have courses whose first year or two is secondary work in everything but name, it will be appreciated that the special agricultural school is not such a new thing in our educational system, and that only as such schools are developed will the state agricultural college be able to serve as a research center and do work of as truly college grade as the other colleges of the state.

One of the questions perplexing the small high school is: How can we relate the teaching of our sciences to agricultural education? The question is largely bound up in the large proposition of making agricultural instruction "incidental," or "correlated" strongly with the other sciences, *versus* the proposition of teaching it entirely separately. The ideal would involve a combination, but in schools teaching agriculture separately, many presenting the subject in but one year, there is a woeful lack of any such tendency. So far as I can discover, the sciences are taught just as abstractly, in most cases, whether agriculture is in the school or not. Or where an attempt at "correlation" is made, it is not correlation but repetition, which may have all the value of a review but none of the charm of new study or new viewpoint.

The desirable kind of correlation is illustrated by a number of examples described in Chapter III. Attention may again be called to a notable instance, that of the work done by the high school at Odell, Ill. No separate instruction in agriculture as such was given. Injurious insects and their relation to birds were studied in zoology. Much of the course in botany took a decidedly practical turn.

The visitor was furnished with essays on "The Corn Plant" that were summaries of various fragmentary studies on the corn stem, under the topic of stems, of the ear and tassel, under flowers and fruits, and in connection with other regular botanical topics. They certainly showed as comprehensive a knowledge of the plant, its habits, and the industry depending on it, as one usually finds where a special text is used. The experimental plat cultivated under the direction of the staff of the agricultural college of the state university is an object lesson for the community rather than for the high school, although the pupils are taken to it for instruction. It furnishes what Professor David Eugene Smith calls "real problems" for the arithmetic classes in the grades. They are taken there to get the data for problems in percentage, linear and surface measure, and other topics.

In the John Swaney School, in Illinois, which offers several courses in industrial subjects, was seen some of the best work done on insects as related to farm industry, not as a part of a definite agricultural study, but as a part of the course in zoology.

The use of the polariscope by the chemistry class at St. Louis, Mich., is an equally good illustration. This instrument is not one of the common instruments in a high-school equipment. But this school is in the heart of the sugar-beet territory. Sugar beets constitute the best paying crop produced, although as yet not more than one acre in ten is put into beets. Several of the boys have gone from the high school into the sugar factories, and have become assistant chemists at a better salary than they could have commanded in the callings usually open in a country town. This school also offers other courses in agriculture.

Likewise the Babcock tester has been used in various physics classes to illustrate centrifugal action.

"Waste of teaching effort" may be well illustrated by a number of instances observed where opportunities similar to those just mentioned had not been used to advantage. A certain well endowed New England academy has a lot of land, a herd of cows, a Babcock tester for their milk, and a cream separator; but no use is made of either piece of apparatus by the physics teacher, who is more intent on the fifty "Harvard experiments," none of which call for the use of those articles. There is an orchard on the school grounds, but the botany teacher makes no use of it, nor does he even show what a graft is, although his class studies the structure of the stem and the cambium cells with the compound miscroscope. A western high school has courses in agriculture taught by a man who had considerable training in the state agricultural college. Some of the students in his class in farm mechanics failed to "pass" because they had not mastered the mechanical principles involved in the machinery. They took their physics under a different teacher, who did not use these implements to illustrate the principles in the mechanics he was teaching as a branch of high-school physics. Most schools doing laboratory work in botany make cross and longitudinal sections of the corn kernel, but they do not test the seed for viability, or test the relation between depth of germ and vigor of growth of the seedling.

Waste of educational energy is a fault that always attends poor correlation or lack of effort altogether in this direction. The example of the class work in physics and in farm mechanics just mentioned is an instance. One school in its printed course of study shows "plant life" as a study in the first year, and botany as a third-year study. The former turned out to be more or less book study of plant functions, and the latter, book study of plant structures, until the spring flowers bloomed, after which it was plant analysis, given for its "disciplinary value," as the principal expressed it. But there was, from the nature of the case, more or less duplication of effort in the two classes, carried on as they were by different teachers.

### *Relation of the Agriculture Taught in the High School to that Taught in the Elementary Schools*

It is a significant fact that the states requiring agriculture to be taught in the rural schools have shown the slowest voluntary development of this subject in the high schools, the only secondary school work having been, till recently, in the technical agricultural schools of Alabama, Georgia, and Wisconsin. Nebraska has the work well developed in the high schools maintaining training classes for teachers, and in some other high schools. Georgia teachers must pass an examination in agriculture for first and second grade certificates. But the subject was not made mandatory before the teachers were partly prepared, at least, to teach the subject. A serious difficulty as yet is the lack of differentiation between agricultural instruction of high-school grade and that of elementary grade. Of several summer schools, some ostensibly train for high-school work, some for normal training schools, others purport to train only for the grades. But so far as could be determined, there was in most of these classes no distinction in the grade of material used or complexity of subject matter to correspond with the difference in the pupils who were finally to receive this instruction. It is as if all of our teachers, high school and elementary, were to get their training together in the normal training classes or in the colleges. The method and subject matter of agricultural instruction must some day become more definitely standardized with reference to elementary and high-school grades. The frequent objections to the text-books used by the high schools reporting indicate that this faulty standardization is recognized in many quarters.

There is as yet no agreement as to the age at which agricultural instruction should begin in a formal manner with profit to the child. The laws of several states require it to be taught in the seventh and eighth grades. The rural common schools, as a rule, use the same texts that the high schools use when carrying on the course for one year or less. If we accept Dewey's definition of education as "a working over of experiences," we must grant that children three or four years apart in age have very different stores of experience, and need different treatment as well as texts. Professor Bailey and others seem to be against agriculture as a study by itself and apart from nature-study until the high school is reached. Professor Stevens, of North Carolina, in the course of nature-study planned for that state, provides for text-book work in agriculture in the fifth grade, to be preceded and followed by work in nature-study. The general trend seems to be toward a specialization in the South and West in the upper grades.

Leaders in the National Society for the Promotion of Industrial Education have much to say just now in regard to the desirability of introducing a differentiation at the age of twelve; but they seem ignorant of the great stride that has already taken place toward that very thing in the industrial education of rural communities. The administrative and legal machinery for it is in much better shape than the methods for carrying it on. The reader will recall the report given in Chapter VI on the grades pupils have or must have completed in order to enter the special schools; that the special agricultural and semi-agricultural schools of the South are settling upon an elementary course of seven grades; that the special schools of the North desire completion of the eighth grade but often make exceptions; that the special schools of Oklahoma require the completion of the eighth grade but provide a preparatory course with elementary agriculture in each of its three years. Added to this is the fact that most of the states provided with these schools require agriculture to be taught in the last of the elementary grades, the seventh or the eighth as the case may be. These various plans actually operate to bring the children into the pre-vocational work at anywhere from 12 to 16 years of age, according to the opportunities of the pupils for completing each year's work, which is admittedly meagre in many rural sections because of the home demands in the spring. So it would seem that the only conditions necessary to make vocational work, or a near approach to it, a vigorous actuality in rural education are money and prepared teachers. The present lack of these seems appalling, but the great strides now being made compare favorably with the progress in the corresponding vocational training in cities.

### Difficulties of Instruction

The most immediate problems have to do with the facilities at the command of the superintendent or principal. The difficulties are many and varied. The responses to the question, "What are your chief difficulties," were numerous and covered a wide range. Although expressed in many ways, they may be roughly grouped under six general headings as shown below:

Table 51

Difficulties Experienced in Teaching Agriculture

| | |
|---|---|
| Lack of equipment and (unspecified) facilities | 69 |
| Lack of time or suitable season for the work | 61 |
| Lack of suitable teacher | 8 |
| Lack of moral support in various forms | 17 |
| Lack of suitable text-book | 9 |
| Difficulties of organization or methods | 21 |
| Total number of difficulties mentioned | 185 |
| Number of schools reporting them | 151 |
| Schools reporting no difficulties | 13 |
| Total number of schools reporting | 164 |

One hundred and nineteen of these were high schools and 45 were training classes. Thirty-eight high schools failed to report that otherwise furnished rather full data. Twenty-eight of these, however, reported that the attitude of the pupils or patrons, or both, was favorable and often enthusiastic. Over half of the 44 training classes failing to report on this topic were in New York.

Some of the variations in the lack of facilities reported, were: lack of apparatus, laboratory space, opportunities for practice, material, fields for observation, grounds for gardens and experimental work; lack of heat in the building over night, reference library. Some complained of the shortness of the growing season before the end of the term, others of the lack of time for field trips. A very small number lamented their own lack of training, although many, no doubt, were conscious of their own lack of preparation. Two objected that the teachers were city girls.

### *The Time Problem*

Superintendents complain of the lack of suitable teachers, and of the unsympathetic attitude of the science teachers, which is serious enough. Such teachers could help wonderfully in solving the time problem. In some cases the botany has been absorbed by the agriculture, with the result that all the necessary botany was given as before, and there was a freedom and elasticity in the year's work that was lacking when the agriculture was supposed to have the right of way for only 12 or 18 weeks. The secondary course in agronomy, outlined in Circular 77 of the Office of Experiment Stations, contains so many topics that are purely botanical, in the sense of being included in all the high-school texts, that it would seem that the course would require to be expanded but little to include about all the botany a high-school course need embrace; or else that the botany should be strengthened in several of its evident weak spots, throwing out certain superfluities, and so include practically all of the projected course in agronomy. In the East and South, more than in the West, botany seems to mean plant analysis; and in order to get any study of plant functions we must apparently inject a new study into the curriculum. For many years the West has included all this in its high-school botany work, due to the influence of Professors Coulter, Barnes, and others, and to the Central Association of Colleges and Secondary Schools. What they have lacked is the agricultural viewpoint rather than new botanical content to make their work symmetrical.

Some of the administrative difficulties mentioned were peculiar to the subject; as, determination of its place in the curriculum, laboratory work hard to organize, no definite outline to work from, or having city children in the school. These difficulties will disappear as theory and practice crystallize, or at least are better understood. Such troubles as lack of practical work, difficulty to get the pupils to do experiments, to observe, to apply the work, and to see that it is real, all these are in a degree dependent on the teacher rather than on the subject, and are difficulties which the same teachers would find with almost any science study.

*The Equipment Problem*

The complaint of insufficient apparatus has less grounds to rest on than would a similar complaint urged for almost any other science, for in no other study than agriculture do home-found appliances come so nearly equalling bought apparatus in serviceableness. Tin cans, perforated and unperforated, paint pails, soup plates, alcohol lamps made of ketchup bottles or even shoe-blacking boxes, are very serviceable. In fact tin cans do better for some experiments than crockery. Lamp chimneys cost little, family scales and spring balances are inexpensive, but chemical thermometers cost more than the ordinary types. With the exception of a Babcock milk tester, four or five dollars should provide all the apparatus needed in a small high school, in addition to what can be made, to perform most of the experiments which the younger high-school pupils can understand. Much of this apparatus should be in the equipment for physics or chemistry if the school offers to teach those subjects at all.[7] In the schools visited there was too much evidence of money invested in showy but almost useless airpumps and static electrical machines while the school suffered from a dearth of simple material or of duplicates of common apparatus necessary to carry on individual laboratory work. The frictional electric machine does not illustrate anything of sufficient commercial importance to justify its cost.

One can well sympathize, however, with those teachers who feel the lack of room, who lack window sills to set plants in, and plain tables on which to spread out corn ears and simple seed testers, or who cannot find a convenient corner in which to keep the soil used for experiments.

---

[7] Attention is called in the bibliography to a number of reports and bulletins containing lists of apparatus suitable for agricultural courses in high schools of various standards.

### *The Teacher Problem*

Few high schools of villages and the poorer townships can aspire to get a teacher trained in agriculture as easily as they now get teachers trained in Latin or mathematics, because the present supply is so much smaller than the number of small schools already teaching the subject and the competition for the men available is too keen on the part of institutions able to pay much larger salaries. Reference to Chapter V will show how large a proportion of the graduates of the agricultural colleges is absorbed by the colleges themselves, the experiment stations, and the state and federal governments. Commercial lines have also attracted a number. A census taken in another year or two would show a large percentage of these graduates secured by the special state and county schools being so rapidly organized. This last source of competition will probably absorb practically all the output of the colleges who have the advantage of teaching experience or pedagogical training. As the salary necessary to secure the desirable men equals or exceeds that paid the principal of the smaller high schools, the only way for such schools to have agriculture taught by a teacher fitted to do so is to elect principals competent to handle the work.

The figures just referred to show the following significant facts:

Of all the teachers of agriculture reported upon, 109 are principals or superintendents, and 51 are assistants, including agriculturists. Sixty-eight of the 79 teachers reported from Missouri and Ohio are principals or superintendents. A majority of the teachers in these two states receive less than $612 and $733 respectively.

The salaries of seven-eighths of the 33 trained agriculturists teaching in secondary schools, who reported their salaries, are $750 or over, including three-fourths of those who are only assistants.

### *The Text-Book Problem*

When asking for a statement of difficulties, no clue was given as to what it was supposed they might or could be; it was quite surprising to find so many respondents characterizing their text as "thin," or "kindergarten." The printed page was evidently not meat but milk for the young mind beginning to realize its own power. Judicious use of government bulletins, which could be obtained free, might have served as an agreeable corrective.

With the appearance of new books which are sufficiently advanced for use in the high schools, the text-book difficulty is solving itself. Books written for the elementary schools are being relegated to use in the grades for which they were intended. In this list may properly be included nearly all of those given in Table 34 as the texts used in schools reporting. Two or three others that have appeared in the last two years, are of the same class. The quality of some of the latest does not promise any contribution to the problem of teaching agriculture in the grades. The texts by Bailey, Jackson and Daugherty, and Ferguson and Lewis, are in many places too difficult for use below the high school. Warren's "Elements of Agriculture" is intended for the upper years of the high school, and seems to be the first book written expressly for that purpose. There seems to be a noticeable tendency on the part of the special secondary schools of agriculture as shown by their catalogues, to make liberal use, as texts, of books that have heretofore found their function in high schools solely as reference books, and that have not been used as texts outside of the agricultural colleges. The result of this on the colleges is obvious.

### *The Methods Problem*

Agriculture is probably taught as well as other sciences in the same schools. But the deficiencies are more prominent on account of the greater opportunity afforded to make concrete the principles of the various sciences. So much have the sciences been regarded as instruments of a disciplinary education, that the absence of concrete applications has not seemed to many to be such a marked defect. The pedagogy of agricultural instruction must take account of the essentially utilitarian aspect of this study. The philosophy underlying the methods of instruction is not consistent with that conception of education, that to be cultural is to be useless; nor does agriculture in the schools depend for its justification on any supposed disciplinary values.

Not that it does not possess as much value in this direction as other studies, but agriculture as a study may justly claim to have a content of its own that is worth while. It does not need the prop of a disciplinary conception of education that bids fair to become obsolete. But if the administrator's idea is to teach the art or trade of farming, his methods, while involving the idea of doing, will probably be those of purely imitative doing, and not be calculated to cultivate initiative, to give opportunities for forming and correcting judgments, nor for acquiring a scientific habit of thought. Viewed as an instrument of education, agriculture should do all these things as truly as any other science is supposed to do. We must remember that we are teaching children as well as subjects.

The meaning may be made more clear by referring again to a stock experiment in agriculture, one that illustrates so many principles of teaching, namely top-grafting. I have mentioned the study of the stem stopping with a microscopic examination of the cellular structure without any attempt to show vividly the function of the cambium layer of cells by having the class make grafts. The practice in teaching agriculture is usually for the teacher to demonstrate the mechanical process without the children knowing much about the structure upon which the success of the experiment depends. They may be told that the scions must be inserted in the cleft at the bark, but I never heard of any one having a trial experiment made of putting one scion in the middle, or heart-wood portion, of the stock in order to demonstrate that only at the outer part of the limb would the knitting together occur. The following illustrates still better how we might, but do not, teach the scientific method of thought. A boy in Ohio studied at a school that was not fortunate in possessing a school garden and he had none at home, but he was anxious to try the efficacy of treating seed potatoes for scab. So he volunteered to demonstrate to a neighbor, whose patch did not produce well on account of scab, the value of the treatment on his lot, and was permitted to do so. The owner of the land was no doubt delighted with the result, and as an object lesson to the landowner it was a good thing, but as a school exercise for mental training it might have given greater returns for the time spent if the teacher had told the boy to leave one-half of the field untreated. It would not have taken one minute longer, it would have taken less time if anything. The crop could have been no worse than it would have been anyway, and all other factors would have been eliminated except that of the special treatment of the potatoes planted. In this way, and only in this way was it possible to show that the treatment, and not favorable weather or extra sprinkling, or non-appearance of bugs, or other causes, was responsible for the improved yield, even in the absence of scab.

In demonstration plats attached to schools this idea of the "control" or "check" experiment is of utmost importance. Otherwise it can not be proved scientifically that the results are not due to superiority of the soil, drainage, or fertilizer used. In schools that had orchards at their command, everything in sight was pruned and no trees exposed to the same conditions were left unchanged, so as to have a basis of comparison in yielding season. The prevailing idea was that of a good workman and not of the investigator. The famous corn plat of the University of Illinois should be commended to all schools teaching agriculture. This corn plat has yielded for the last three years an average of 27 bushels to the acre, while another plat near it yielded, under a different system of farming, at the rate of 96 bushels. Farmers who visit the experimental farm show considerable contempt for this field until they learn that it is an object lesson on how *not* to do it, and that it has taken about thirty years to get this field in its present poor condition by keeping it in corn.

We do not have to believe that the unrelated chemistry experiment is the only thing giving opportunity for making and correcting judgments; nor is this the exclusive attribute of that particular kind of mathematical physics that is killing itself off except as bolstered up by college entrance requirements. The contests between the disciplinarians and the phenomenologists tend to drive the latter class into an extreme and untenable position. The remark made recently that "there are no methods of teaching above the grades" is an indictment of the high-school instruction and not of pedagogy. And even in the grades the current methods of carrying on garden work are not calculated to encourage much initiative on the part of the child or to place any definite problems before him for solution. It is often only a sort of physical exercise that is better than gymnasium work because it is out of doors.

### *The Attitude of Students and Patrons*

Much of the success of the instruction depends on the attitude of the pupils, and the encouragement given the work by the parents. On the other hand, the degree of interest shown reflects in no small degree the quality of the teaching. Opinions were asked regarding the attitude of the students and patrons, and are here given for what they are worth. Attention is called to various possible factors operating against the value of these judgments, and to possible explanations of some of the unfavorable cases. Very few, only ten each, of the high school and normal training classes, failed to offer some estimate. The answers are classified rather arbitrarily according to the degree of interest indicated with the most frequently recurring phrases indicated. The opinions of the high schools are separated from those of the normal training classes. The first number given refers to the former and the second figure to the latter.

TABLE 52

THE ATTITUDE OF PUPILS AND PATRONS TOWARD AGRICULTURE

| | Reported by high schools | Reported by training classes | Total |
|---|---|---|---|
| Pupils "enthusiastic," "very much interested," "very favorable," or study "very popular" | 30 | 37 | 67 |
| Pupils "like it," "pleased with it," "take kindly to it," or attitude "good," "favorable," "pleasing" | 90 | 66 | 156 |
| Attitude "fairly good," "tolerant," pupils "show no marked interest" | 12 | 7 | 19 |
| "Indifferent," "backward," "no interest" | 4 | 5 | 9 |
| "Unfavorable," or pupils "dislike it" | 3 | 1 | 4 |
| "Cannot judge," or "study too new" | 2 | 2 | 4 |
| Not reporting | 10 | 10 | 20 |
| Total | 151 | 128 | 279 |

A few of the extreme expressions may be of interest: "Pupils enjoy the work beyond all my expectations," "heart and soul in the work," "like it, the only trouble there is not enough." Concerning the attitude of the patrons in particular: "Would not do without it," "watch the work closely," "regard the work as practical," "parents read the text-book." One superintendent said that his class could not get the books because the farmers bought them as soon as they arrived in town, and that the bookseller had to order the books three times. In two or three cases the pupils were reported as interested, while the patrons were neutral or hostile.

In nearly every case reporting a dislike for the subject, the returns also showed that the work included neither class-room experiments, demonstrations, nor practical home work. Under the circumstances one would expect nothing else. Others reported "no demand for it," "prejudice against book farming," and one, that it was hard to make the work seem practical to the patrons. Many of these cases, I believe, could have been managed by not labelling the study so conspicuously as something never before taught, and, by incorporating instead, the material in subjects already on a safe footing. In one school the work was introduced under the heading of geology to meet the objections of an influential citizen. He has since given several hundred dollars for agriculture, and it is taught in all four years of the high school. In contrast with the cases of indifference just noted, is the school whose superintendent told me of knowing that, out of twenty graduates that year, four would not have entered the high school from the country but for the respect their parents entertained for the unpretentious course in agriculture. Another school placed animal husbandry in the first year, an unusual place, because that work appealed to the parents as so eminently practical that they were willing to have the boys continue in school in order to get that kind of work. Two city boys graduating from a school where the subject was regarded "with doubt" chose to enter the state agricultural college.

The pupils in a New York high school are "indifferent because the work is too closely allied to their home life (!); they want something new, and seek to avoid farm life." On the other hand, the pupils of a Nebraska school "like it because most of them are rural pupils." The former school reports no experimental work while the latter school does. In another school "it is hard to get the pupils interested at first because most of them think they know all about it."

The fact must not be overlooked that enthusiasm on the part of the teacher may be reflected not only in the attitude of his pupils but sometimes in his notion of their feelings, and thus give rise to a roseate but unjustified answer. Again some teachers might fear that any but a favorable report would reflect discredit on their work. Then, too, there is to be considered that familiar tendency to give, however honestly and unintentionally, the kind of an answer one thinks the inquirer would like to get.

However, most answers that would fall under any of the above criticisms are, perhaps, counterbalanced by the reports of teachers not in sympathy with a study they must teach against their will. A former superintendent in a small Indiana village wrote that he had not been in sympathy with the movement. His conception of education was that "life was more than meat and the body more than raiment." But the demand for instruction in agriculture was so insistent, so sincere, and so dignified and reasonable, that it could not be ignored. So he planned to put a course into operation but himself "abandoned the field of general school work for the more congenial field of history and psychology."

If it be true that the life in the school should be as little unlike the life outside as possible, as educators of note maintain, or, stated more positively, that the work of the school should be related as nearly as possible to the outside life, it must be especially true of any industrial phase of the school life. It is incumbent on the agriculture taught in the high school to be particularly relevant to the principal activities of the immediate neighborhood and to give an insight into the importance, if not the methods, of agricultural interests in other parts of our nation. Agriculture in the broad sense includes a variety of activities. More of these are represented in some localities than in others. Some are much more widespread than others.

An effort was made in this investigation to learn the important local industries that "fit in" well with the agriculture of the school and that are interesting to the pupils. While a detailed classification would include more than thirty headings, the answers may be roughly grouped under the seven given below. Those from the high schools are kept separate from those of the normal training classes as in the last table.

TABLE 53

PRINCIPAL INDUSTRIES OF THE COMMUNITIES SUPPORTING AGRICULTURAL COURSES

| | Reported by high schools | Reported by training schools | Total |
|---|---|---|---|
| General farming and farm crops | 90 | 12 | 102 |
| Special farming and farm crops | 10 | 6 | 16 |
| Horticulture and gardening | 21 | 10 | 31 |
| Animal husbandry, including dairying | 68 | 25 | 93 |
| Manufactured milk products | 6 | 7 | 13 |
| Manufactures allied to agriculture | 6 | 8 | 14 |
| Industries not allied to agriculture | 3 | 3 | 6 |
| Reporting "no local industries" | 6 | 4 | 10 |
| Total | 210 | 75 | 285 |
| Number of schools represented | 147 | 44 | 191 |
| Number not reporting on blanks returned | 23 | 31 | 54 |

(Twenty of the thirty-one training schools not mentioning this point were New York schools.)

One of the most important of the teaching problems concerns the relative degree of difficulty of the various topics usually presented to high-school pupils. The correspondents were asked to name the agricultural topics giving them the most concern. It is a matter of regret that the number of responses on this point is so much smaller than on most others. The following replies represent 55 high schools. The few who complained of the difficulty of teaching anything requiring laboratory work may be dismissed from further consideration. For convenience, the points are arranged under four general headings.

TABLE 54

TOPICS IN AGRICULTURE MOST DIFFICULT TO TEACH

| Soil work: | No. | Animal husbandry: | No. | Plants and crops: | No. |
|---|---|---|---|---|---|
| Soils | 17 | Feeds and feeding | 14 | Field crops | 2 |
| Soils and rocks | 2 | Feeds, analysis of | 1 | Horticulture | 1 |
| Analysis of | 3 | Animal husbandry | 1 | Orchards | 1 |
| Chemistry of | 8 | Dairying | 1 | Plant breeding | 2 |
| Drainage of | 1 | Livestock | 1 | Plant propagation and improvement | 1 |
| Fertility of | 4 | Stock raising | 1 | Budding and grafting | 1 |
| Nitrification of | 1 | Total | 19 | Plant diseases | 2 |
| Physics of | 3 | Miscellaneous: | | Chemistry of plants | 1 |
| Fertilizers | 2 | Insects | 2 | Total | 11 |
| Fertilizers, artificial | 1 | Pests | 1 | | |
| Tests | 1 | "All tests" | 1 | | |
| Total | 43 | Total | 4 | | |

| | |
|---|---|
| Total of items | 74 |
| Number of schools reporting them | 52 |
| Number of schools reporting "none" | 4 |

One report stated that all the topics were easy, another that all were of equal difficulty, and a third that all were too easy in the text. Eight of the schools reporting difficulties teach the subject in the third year, and four in the fourth year.

Many, no doubt, failed to report because no one topic stood out prominently as being particularly more difficult than the others. The concentration of the replies on two points, soils and feeds, may mean (1) that these two subjects are not treated in the texts with the same clearness as the others; (2) that they are inherently more difficult than other topics; or (3) that they are, in many respects, too difficult for the pupils in the lower years of the high school. Probably all three factors are concerned. The writer has called attention at some length[8] to the uneven and otherwise unsatisfactory treatment of "soils" in books of high-school grade. Justification of the second suggestion is found in the fact that both topics draw heavily on

[8] Some Text-Books for Secondary-school Agriculture, *Nature-Study Review*, Vol. III, pp. 180–185.

physics and chemistry, studies usually deferred until the third and fourth years, and in the fact that the large proportion of difficulties on this point did not appear in the reports from schools teaching agriculture in the last two years of the course.

The fact has not been sufficiently recognized that the topics in agriculture, and the treatment of those topics, must be as carefully graded as the subject matter in any other branch of knowledge. One principal stated, wisely it would seem, that he attempted to handle but three lines, and found them not too difficult for his pupils. A proper organization of courses will eliminate certain topics from the work of the first years, or treat them in a more elementary way. Where the simpler treatment of soils, for instance, is given in the last grade of the elementary school, it is manifestly unwise to repeat the work, even in a more "advanced" manner, when it has been impossible as yet for the student to gain the scientific background for more advanced work. If a school can include the subject of agriculture formally in but one year, these more difficult topics might much better be deferred for treatment in connection with the sciences of the upper years. We may yet see high-school texts, or series of texts, written in "parts," these parts one, two, three, etc., not treating with completeness the various departments of agriculture, but containing work appropriate to different grades, as the seventh, eighth, tenth, and twelfth or eighth, ninth, and twelfth. One of the encouraging signs is that men who are not ostensibly writing texts, are rendering invaluable aid by issuing in small compass guides for practical work in several restricted fields of agricultural instruction.

### *Help That May be Given the Schools*

High schools are greatly in need of aid that can only be given by agricultural colleges, experiment stations, and departments of agriculture. It is almost impossible to obtain anywhere at a reasonable cost small working collections of insects of economic value, of weed seeds, and of soil-forming rocks. An occasional college can be found that has, at one time or another, provided such collections, but their willingness has usually outstripped their appropriations. The United States Geological Survey once supplied free of cost an exceedingly good collection of common rocks, although a nominal price would have been justified. Few schools know where to send for graded samples of grains and reliable samples of fertilizers. The many bulletins and reports issued by the federal government and various states are invaluable for reference, but there is still great need for studies worked out by the state colleges or still better by trained agriculturists now in secondary work, adapted to the local needs of different sections of the country.

## Conclusions and Summary

A wise and far-reaching policy regarding agricultural education in the public school system is highly desirable, both for the sake of the efficiency of the work itself, and for the protection of our present high schools.

Legislative aid is beneficial; legislative mandates are of doubtful help.

Local high schools can not hope to do as pretentious work as the special schools because of the excessive cost of a large plant.

The special school may be able to work out cultural material suitable for rural students better than the present high schools.

The special school may become either a competitor of the present rural high schools or an adjunct to the system we now have in the wealthier states, crippling them in one case and stimulating them in the other.

Such available opinions as are based on personal experience do not warrant the fear of the evil influence of the small villages in which the special schools so far have been located, when they are not in the country altogether.

The attitude of the colleges and universities toward the agriculture taught in the high school is as favorable as the character of the work merits.

Agriculture in the schools is a very unsettled and undetermined thing.

Agricultural instruction must be adapted to the community. A general course will fit but very few places.

We need clearer ideas regarding the pedagogical principles involved in this and other science teaching.

We need a clearer understanding regarding the domain of this and the other sciences, and we also need more coöperation between them to save time and teaching energy. The sciences should change their viewpoint, and thus allow agriculture to put its time on the more technical phases of its subject matter.

Agriculture should be recognized as an instrument of education in the sense of affording mental training, as well as in the sense of furnishing an acquaintance with the environment, and should be used as such. The scientific method of thought should have a more definite place in the instruction.

Agriculture is probably as well taught as the other sciences in the same schools.

Current practice places agriculture in the lower years of the high-school curriculum; and so it will probably function more and more as an "introductory science," as physical geography was once expected to do. It can not then avail itself of the training and information gained from the other sciences as some of its advocates would have it do.

As given in the grades, it is very imperfectly differentiated from that of the high school.

This differentiation must largely be worked out and be made apparent by the schools that train teachers for the elementary and high schools.

Agriculture as a separate branch in the elementary-school curriculum is bound up with the question of differentiating our entire system of education at about the age of twelve, as urged by many interested in industrial education.

Lack of equipment need not discourage teachers as it does. We have not by any means exhausted present resources.

Home garden work has proved to be an invaluable aid to school work, and has certain advantages over the school garden.

Lack of time is a matter of will, management, and public opinion. The sentiment of the community is very often ahead of the preparation of the teacher, and is willing to spare time from some of the traditional studies whose chief justification is a supposed disciplinary value.

We need texts in greater variety and better adapted to high-school pupils.

Trained teachers are scarce and command a higher salary than the rural high school thinks it can afford to pay. We must depend largely on giving additional training to teachers already well grounded in science, or with practical farm experience—preferably both.

Any one responsible for agricultural instruction must have worked out an aim and a philosophy underlying it, and must keep in mind the child, the equipment, and the relation to the community.

The attitude of the patrons is usually favorable and the interest of the pupils is related, about as one would expect, to the amount of laboratory work, i. e., to the quality of instruction.

## APPENDIX A

## LEGISLATION PERTAINING TO AGRICULTURAL INSTRUCTION IN PUBLIC HIGH SCHOOLS

An examination of the school laws of a large number of states, most of them as late as 1907 or 1908, fails to show many specific references to the teaching of agriculture in the general public high school. A few important laws on this subject were passed during the sessions of 1908 and 1909.

In Kansas, county high schools may be established subject to certain restrictions, among others that "their course of study shall be four years in length, and shall be such as will prepare for entrance to the freshman year of the college of liberal arts of the state university, of the agricultural college, or to the professional course of the normal school."[1] It would seem that the requirement to teach agricultural branches in the county high schools depends on the entrance requirements established by the agricultural college, unless the law be construed to mean that the agricultural college must take the students who have pursued the course taught in schools that "use the course of study laid down by the state board." (Sec. 185.)

In Maine we find that "the course of study in the free high schools shall embrace the ordinary English academic studies which are taught in secondary schools, especially the natural sciences in their application to mechanics, manufacture, and agriculture. . . ."[2] But so far as learned, nothing pertaining to agriculture is taught in a way to make the relation with the natural sciences apparent. The legislature in 1907 made an appropriation of $500 a year to high schools and academies that would put in a course in agriculture.[3] Several

[1] Laws Relating to the Common Schools of Kansas, 1907, Sec. 177.
[2] Laws of Maine Relating to Public Schools, 1905, p. 21.
[3] Chap. 78, March 20, 1907, amending Sec. 6, Chap. 148, Acts 1901. (Sec. 81, Chap. 15, Maine Revised Statutes, 1903.)

of them started at once to introduce some work in order to get this state aid, but the state superintendent of public instruction ruled that the conditions necessary to receive the aid would not be considered as having been complied with unless an agricultural college graduate, or some one else equally fitted, were engaged to carry on the work.

Michigan provides more specifically for agricultural instruction in the township rural high schools, sanctioned by Act 144, 1901, which provides that "the board shall have power: . . . to provide a course of study which shall be approved by the superintendent of public instruction and the president of the Michigan Agricultural College, and shall not consist of more than four years' work. Said course of study may include instruction in manual training, domestic science, nature-study and the elements of agriculture."[4]

The most recent and apparently most effective legislation in Minnesota is the act of the legislature of 1909 from which the following quotations are taken:

To provide for the establishment and maintenance of departments of agriculture, manual training, and domestic economy in state high, graded, and consolidated schools, and to authorize rural schools to become associated with such state, graded, or high schools and making appropriations therefor.

SEC. 1. Any state high school, graded, or consolidated rural school having satisfactory rooms and equipment and having shown itself fitted by location and otherwise to do agricultural work, may, upon application to the state high school board, be designated to maintain an agricultural department.

SEC. 2. Each of such schools shall employ trained instructors in agriculture, manual training and domestic science (including cooking and sewing), and have connected therewith a tract of land suitable for a garden and purposes of experiment and demonstration, containing not less than 5 acres, and located within 2 miles of said buildings or within the school district.

SEC. 3 provides that instruction in the industrial department shall be free to all residents in the state, thus not restricting its use to pupils from the district that helps to maintain the school. It provides also for short courses in the winter months and enumerates a wide range of topics that shall be treated.

[4] [Michigan] Public Laws, Act 144, 1901, Sec. 4, Clause (*g*).

SEC. 4. Each of said schools shall receive state aid equal to two-thirds (⅔) of the amount actually expended upon such departments and vouched for, but in no case to exceed two thousand five hundred dollars ($2,500) per year. Not more than ten schools shall be aided the first year nor more than ten added to the list every two years thereafter. The special aid provided under this act shall be in lieu of all other aid for industrial training granted by the state to schools operating hereunder.

SECS. 6, 7, 8, 9, 10, and 11, provide a way for rural schools to consolidate or to attach themselves to a graded or state high school maintaining such an industrial department, such school then being known as a "central school."

The Nebraska law approved April 5, 1907, provides that in county high schools "there shall be taught and practiced in the ninth and tenth grades, manual training, domestic science, and the elements of agriculture and in the eleventh and twelfth grades normal training and the theory and practice of agriculture for the purpose of teaching and practice. The board is hereby authorized to purchase the necessary apparatus and materials for this purpose, together with a tract of land not less than 5 acres, conveniently situated to said county school for actual practice by all the students or a part of the students under the direction of a competent instructor for experimentation in all forms of agriculture."[5]

The county board of commissioner supervisors constitutes the board of trustees of such county high schools, and tuition is free to all pupils residing in the county.[6]

In Oklahoma the legislation which became effective May 20, 1908, is the most elaborate yet enacted for the promotion of agricultural instruction. While covering all fields of educational activity in the state, the nearest approach to a mention of agriculture in the general high school is the phrase "the public schools."

*Pennsylvania.*[7]—Every high school receiving aid from the state "shall employ for said high school at least one teacher legally certified to teach . . . chemistry, including chem-

[5] School Laws of Nebraska (Elliott Digest), 1907, p. 43, (punctuated as there printed).

[6] Chap. 122, April 5, 1907.

[7] The Common School Laws of Pennsylvania, 1907, Sec. CXII, p. 54.

istry of soils . . ." (Twelve other subjects are also enumerated.)

The legislature of Texas in 1909 passed the following law providing state aid for establishing departments of agriculture, but expressly forbidding aid in maintaining them after the first year:

It shall be the duty of the state board of education to duplicate by an appropriation out of money provided by this act, any amount not less than five hundred dollars ($500) and not more than two thousand dollars ($2,000) that shall have been appropriated and set apart by the trustees of any common school district for the purpose of establishing, equipping, and maintaining departments in their respective schools for giving instruction in agriculture, including such courses in manual training and domestic economy as are subsidiary to agriculture; *provided,* such appropriation or donation shall not be made more than twice to the same school, and provided that in granting such appropriations to high schools the state board of education shall consider the geographical location of the school applying, with a view of locating if possible one school in each of the senatorial districts of the state. The board of trustees of a school seeking aid in establishing, equipping, and maintaining in their high schools a department for the teaching of agriculture, including such courses in manual training and domestic economy as are subsidiary to agriculture, shall provide ample room and laboratories for instruction in botany, zoology, and such other elementary sciences as are necessary to instruction in secondary agriculture, and shall provide a tract of land conveniently located, which shall be sufficiently large and well adapted to the production of farm and garden plants, and shall employ a teacher who has received special training in agriculture and allied branches. The state superintendent of public instruction shall make full and accurate investigation of the school property, appliances, and ground possessed by any board of trustees that may seek aid under the provisions of this act, and he shall also inquire into the qualifications of the teacher or teachers who are to give instruction in agriculture, manual training, and domestic economy in the school or schools seeking aid under the provisions of this act, and shall make a report of the result of his investigation to the state board of education, together with his conclusions and recommendations touching the same. The state board of education shall grant aid to those high schools that have complied with the provisions of this act and that have been recommended by the state superintendent of public instruction and that shall give evidence that after the state aid is withdrawn the district will continue to maintain the department for instruction in agriculture out of its own funds.[8]

Section 6 of the same Act says:

The sum of thirty-two thousand dollars ($32,000) or such part thereof as is necessary is hereby appropriated out of any money in the state treasury, not otherwise appropriated, for the year ending August 31, 1910, and thirty-two thousand dollars ($32,000) or such part thereof as is necessary for the year ending August 31, 1911, for the purpose of carrying out the provisions of article 5 of this act.

Vermont has the following indefinite provision: ". . . and instruction may be given in political, social, moral, and industrial sciences . . . ,"[9] although no high school yet teaches anything relating to the farming industry. However, the principal of Vermont Academy, at Saxton's River, conducted a small but enthusiastic class in agriculture in the year 1906-1907.

Virginia has among other "requirements necessary to receive aid from the state high school fund" the following: "The course of study shall conform to the outline given herewith," which provides for elementary agriculture and lessons in botany and zoology for three periods a week through the second year, elements of physics and elementary agriculture or elementary chemistry and elementary agriculture for three periods a week through the third year, and the elements of chemistry or the science of agriculture for three periods a week through the fourth year.[10]

The Virginia legislature has appropriated "for high schools to be expended as per act creating public high schools, one hundred thousand dollars, provided that so much of the five hundred and seventy-five thousand dollars herein provided, for the support of the public and high schools as may be necessary, not to exceed twenty thousand dollars, shall be devoted to the establishment of departments of agriculture, domestic economy, and manual training, in at least one high school in each congressional

---

[8] School Laws of Texas, Sec. 12, p. 8. Acts of 31st Legislature, Chap 113, Sec. 5.

[9] General Laws of the State of Vermont Relating to Public Instruction, 1907, Sec. 1016.

[10] Standard of Requirements for High Schools [Virginia], Board of Education, 1906.

district of the state, to be conducted under such rules and regulations as the state board of education and the president of Virginia College of Agriculture and Polytechnic Institute may prescribe."[11]

[11] Statutes of Virginia, Acts of 1908, p. 420.

# APPENDIX B

## LIST OF REFERENCES ON AGRICULTURAL EDUCATION

NOTE. The following list contains references published, for the most part, during the last two or three years. In but few instances does it duplicate references given in the bibliographies attached to the studies by Dr. Jewell and by Professor Bailey, cited below. References to text-books, catalogues and publications dealing primarily with work of the elementary grades have been omitted. These references have been classified topically by number on page 200.

1. ABBEY, M. J. Normal School Instruction in Agriculture. Office of Experiment Stations, circular no. 90, Dec. 16, 1909, pp. 31.

2. ANDERSON, LEROY. Agriculture in the High Schools. University of California, College of Agriculture, circular no. 47, Nov., 1909, pp. 18.

3. BABCOCK, E. B. Agriculture in Secondary Schools. *Nature-Study Review,* vol. 5, no. 8, Nov., 1909, pp. 210-218.

4. BAILEY, LIBERTY HYDE. On the Training of Persons to Teach Agriculture in the Public Schools. Bureau of Education, bulletin no. 1, 1908, pp. 52.

4a. ———. Education by means of Agriculture *in his* Cyclopedia of American Agriculture, Vol. IV, chapter VIII, pp. 355-477.
Gives the best general treatment extant of the early history of the movement.

5. BALCOMB, E. E. Some Means of Awakening and Maintaining an Interest in Agricultural Education. National Education Association. Journal of proceedings and addresses, 1909, pp. 959-963.

6. BARTO, D. O. The Preparation of Teachers of Agriculture for the Public Schools of Illinois. pp. 8. Published by the University of Illinois.

7. BISHOP, E. C. The Present Status of Agricultural Education in the Public Schools. National Education Association. Journal of proceedings and addresses, 1909, pp. 976-982.

8. BRICKER, GEORGE A. Elementary Agriculture in the New Holland High School. Ohio State University, Agricultural College, Extension Bulletin, vol. 3, no. 7, March 1908, pp. 3-6.

8a. ———. The Teaching of Agriculture in the High School, pp. 202, New York, 1910.

9. BROWN, ELMER ELLSWORTH. Making of Our Middle Schools. pp. 335 *et seq.*

10. ———. Some Notes on Agricultural Education. National Education Association. Journal of proceedings and addresses, 1908, pp. 1199-1202.

11. BUTTERFIELD, KENYON L. Agricultural Education in the Schools. Transactions of the Massachusetts Horticultural Society for 1908, I, pp. 111-121.

12. CALDWELL, OTIS W. The Course in General Elementary Science for the First Year of the High School. Proceedings of the Ninth Meeting of the Central Association of Science and Mathematics Teachers, 1909, pp. 115-127.

13. CLUTE, W. N. What One Class in Agronomy Did. *School Science and Mathematics,* vol. 9, no. 8, Nov. 1909, pp. 731-735.

14. COOK, G. B. The Agricultural and Industrial Educational Movement in the South. Proceedings of the Conference on Education in the South, 1909, pp. 69-84.

15. COOK, J. W. Progress of Education for the Year. National Education Association. Journal of proceedings and addresses, 1909, pp. 390-397.

16. COOK, O. F. Agriculture the Basis of Education. *Monist,* July 1907, pp. 347-364. [Reprint.]

16a. CROCHERON, B. H. Laboratory and Field Work in the Agricultural High School. National Education Association. Journal of proceedings and addresses, 1910, pp. 1089-1093.

17. CROSBY, DICK J. Progress in Agricultural Education, 1906. Annual Report of the Office of Experiment Stations, 1906, pp. 213-300. [Reprint.]

18. ———. Progress in Agricultural Education, 1907. Annual Report of the Office of Experiment Stations, 1907, pp. 237-306. [Reprint.]

19. ———. Progress in Agricultural Education, 1908. Annual report of the Office of Experiment Stations, 1908, pp. 231-288. [Reprint.]

19a. ———. Progress in Agricultural Education, 1909. Annual report of the Office of Experiment Stations, 1909, pp. 251-325. [Reprint.]

20. ———. Special Agricultural High Schools. National Education Association. Journal of proceedings and addresses, 1909, pp. 974-976.

20a. ———. The Place of the Agricultural High School in the System of Public Education. National Education Association. Journal of proceedings and addresses, 1910, pp. 1103-1107.

21. ———. The Use of Illustrative Material in Teaching Agriculture in Rural Schools. Yearbook of Department of Agriculture, 1905, pp. 257-274. [Reprint.]

22. ———. The Work of the National Government in Extending Agricultural Education through Public Schools. National Education Association. Journal of proceedings and addresses, 1907, pp. 1063-1069.

22a. Cyclopedia of Education. *Article on* Agricultural Education *in* Vol. I, pp. 58-68.

23. DAVENPORT, EUGENE. Education for Efficiency, pp. 184, Boston, 1909.

The most elaborate discussion yet presented on the theory and bearing of agriculture on secondary education.

24. DAVIS, BENJAMIN MARSHALL. Agricultural Education: State Normal Schools. *Elementary School Teacher,* vol. 10, no. 8, April 1910, pp. 376-387.

25. ———. Agricultural Education: United States Bureau of Education. State Departments of Education. State legislation. *Elementary School Teacher,* vol. 10, no. 4, Dec. 1909, pp. 163-176.

26. ———. Agricultural Education: The United States Department of Agriculture. *Elementary School Teacher,* Nov. 1909, pp. 101-109.

These articles by Dr. Davis present a concise summary of the entire field. They have since been issued with others in book form by the University of Chicago Press under the title of Recent Developments in Agricultural Education.

27. ———. What Constitutes Successful Work in Agriculture in Rural Schools? National Education Association. Journal of proceedings and addresses, 1908, pp. 1188-1194.

A questionaire study.

28. ———. *In the* Report of the United States Commissioner of Education for the year 1909, vol. 1, pp. 142-143, on agricultural teaching in California.

29. DRAPER, ANDREW S. Agriculture and Its Educational Needs. *The School Bulletin and New York Educational Journal,* vol. 35, no. 5, Jan. 1909, pp. 84-90. [Issued in book form.]

30. DUNTON, A. M. Agriculture in Rural Schools. *Northwestern Agriculturist,* vol. 24 (1909), no. 6, pp. 7, 8; and no. 7, p. 3.

31. FLETCHER, S. W. An Elementary Course in Horticulture for the Schools of Michigan. Published by the state superintendent of public instruction, bulletin no. 28, 1908, pp. 31.

32. FOWLER, FREDERICK H. Early Agricultural Education in Massachusetts. 54th annual report of the Massachusetts State Board of Agriculture, pp. 331-392. [Reprinted as Public Document no. 4.]

33. French, W. H. Agriculture in the High Schools of Michigan. Michigan Agricultural College, Department of Agricultural Education, bulletin no. 2, pp. 12.

33a. ———. Report of the Michigan State Commission on Industrial and Agricultural Education. Lansing, Dec. 1910, pp. 94.

34. Georgia. District Agricultural Schools of Georgia. Bulletin of the University of Georgia, vol. 7, no. 11, July 1907, supplement, pp. 47.

35. Giles, F. M. The Teaching of Agriculture in the High School. *School Review,* vol. 17, no. 3, 1909, pp. 154-165.

36. Graham, A. B. Agriculture in High Schools. *Nature-Study Review,* vol. 4, no. 3, March 1908, pp. 65-70.

37. Green, Samuel B. Course in Fruit Growing for Movable Schools of Agriculture. Office of Experiment Stations, bulletin no. 178, pp. 100.

38. Guss, R. W. Physics and Agriculture. Proceedings of the conference on agricultural science. Amherst, Mass., July 1908, pp. 29-34.

39. Hamilton, John. Form or Organization for Movable Schools of Agriculture. Office of Experiment Stations, circular no. 79, 1908, pp. 8.

40. ———. History of Farmers' Institutes in the United States. Office of Experiment Stations, bulletin no. 174, pp. 96.

41. Harbourt, S. A. Agriculture in the High School. *Journal of Education* [Boston], vol. 70, no. 16, 1909, pp. 430-431.

42. Hart, W. R. Educational Agriculture, pp. 16.

43. ———. The Place and Function of Agriculture in the Curriculum. *Nature-Study Review,* vol. 5, no. 6, 1909, pp. 161-164.

44. ———. Science Teaching in Some of the Smaller High Schools: *in the* Seventy-first annual report of the Massachusetts Board of Education. [Reprint 5 pages.] (Bound with 67, on page 196.)

45. Haskell, Sidney B. Relation of the Physical Sciences to Agriculture. Proceedings of the second annual conference on agriculture science. Amherst, Mass., July 1909, pp. 44-49.

46. Hays, Willet M. Our Farmer Youth and the Public Schools. *American Monthly Review of Reviews,* Oct. 1903, pp. 449-455.

47. ———. Agriculture, Industries, and Home Economics in Our Public Schools. National Education Association. Journal of proceedings and addresses, 1908, pp. 177-190.

48. ———. Country Life Education. Address before the Pennsylvania State Board of Agriculture, Jan. 23, 1907. Office of Experiment Stations, circular no. 73, pp. 13.

49. ———. History of Secondary Agricultural Education. Proceedings of the twenty-eighth annual meeting of the Society for the promotion of Agricultural Science, 1907, pp. 73-83.

50. ———. Education for Country Life. Department of Agriculture, Office of Experiment Stations, circular no. 84, Aug. 23, 1909, pp. 40.

51. Howe, F. W. School Agriculture in Its Relation to the Community. Rural Life Conference, 1909, Charlottesville, Va., pp. 66-71.

51a. ———. Agriculture as First Year Science. [Editorial in] *Office of Experiment Stations Record,* vol. 23, no. 3, pp. 201-209, Sept. 1910.

52. Hunt, T. F. Agriculture in Secondary Schools. Annual Report of the Pennsylvania Department of Agriculture, vol. 13, 1907, pp. 382-395.

53. James, Hon. C C. Deputy Minister of Agriculture for Ontario. Teaching of the Elements of Agriculture in the Common Schools. Address delivered at the meeting of the Farmers' National Congress, at Boston, Oct. 5, 1899, and printed by the Massachusetts Board of Agriculture, pp. 15.

54. ———. The Teaching of Agriculture. Annual report of the Ontario Agricultural and Experiment Union, 1907, pp. 75-94.

55. Jeffery, Joseph A. An Elementary Laboratory Study in Crops, for the Schools of Michigan. Published by the state superintendent of public instruction, bulletin no. 26, 1907, pp. 28.

56. ———. An Elementary Laboratory Study in Soils, for the Schools of Michigan. Published by the state superintendent of public instruction, bulletin no. 27, 1908, pp. 36.

57. Jewell, James Ralph. Agricultural Education. Bureau of Education, bulletin no. 2, 1907, pp. 140.

Has a good general view of the work in foreign countries.

58. Johnson, Riley O. One Hundred Experiments in Elementary Agriculture for California Schools. State Normal School, Chico, Cal., 1909, pp. 42.

59. Kays, Victor L. The John Swaney School. *Nature-Study Review,* vol. 4, no. 9, Dec. 1908, pp. 271-275.

60. Kern, O. J. The Consolidated School and the New Agriculture. National Education Association. Journal of proceedings and addresses, 1907, pp. 277-279.

61. Main, Josiah. A Manual for High Schools, with Reference to Science and Agriculture. University of Tennessee, 1909, pp. 32.

61a. ———. Educational Agriculture. Western State Normal [Quarterly, Hays, Kansas.] vol. 2, no. 3, pp. 74, Sept. 1910.

62. ———. Agriculture in the High School. *School Science and Mathematics,* vol. 10, no. 3, March 1910, pp. 217-228.

63. ———. The Correlation of High-School Science and Agriculture. National Education Association. Journal of proceedings and addresses, 1909, pp. 983-987.

64. ———. University Extension in Tennessee High Schools. *The School Review,* vol. 18, no. 1, Jan. 1910, pp. 29-35.

65. MASSACHUSETTS AGRICULTURAL COLLEGE. Department of Agricultural Education. Proceedings of the Conference on Agricultural Science in the Summer School of Agriculture, Amherst, Mass., July 1908, pp. 43. Press of Carpenter and Morehouse, Amherst, Mass., 1908.

66. ———. Proceedings of the Second Annual Conference on Agritural Science in the Summer School of Agriculture, Amherst, Mass., July 1909, pp. 58. Press of Carpenter and Morehouse, Amherst, Mass., 1910.

67. MASSACHUSETTS. Public School Agriculture. Report of committee appointed at conference on Agricultural Science, Amherst, Mass., 1908, pp. 38. Massachusetts Agricultural College, 1909.

67a. Massachusetts. Report of the Board of Education on Agricultural Education. January 1, 1911, pp. 97.

A very good discussion from a professedly local viewpoint, but presenting an interesting constructive policy for work in secondary schools.

68. MASSACHUSETTS COUNCIL OF EDUCATION. The Relation of the Massachusetts High School to Community Needs, with Special Reference to the Demand for So-Called Practical Subjects; abstract from a report of a special committee, *in the* seventy-first annual report of the Massachusetts Board of Education. [30 p. reprint.]

69. McDONALD, W. Agricultural Education in America. pp. vii and 162, New York, 1909.

General Survey of the U. S. Department of Agriculture, endowments, farmers' institutes, and agricultural education in Minnesota.

70. MICHIGAN. Department of Agricultural Education. Michigan Agricultural College. A Course in Agriculture for the High Schools of Michigan, 1908, pp. 36; 1909, pp. 48.

71. MICHIGAN. State Superintendent of Public Instruction. County Schools of Agriculture in Michigan, Bulletin no. 24, 1907, pp. 11.

72. ———. Township Rural High Schools in Michigan, Bulletin no. 25, 1907, pp. 17.

73. NATIONAL EDUCATION ASSOCIATION. Report of the committee on industrial education in schools for rural communities, to the National council of education. [Winona, Minn.] Published by the association, 1905, pp. 97.

74. ———. Report of the committee on industrial education in schools for rural communities, to the National council of education, 1907. Journal of proceedings and addresses, 1907, pp. 409-454.

74a. ———. Report of the committee on encouraging college entrance credit in high school agriculture. Journal of proceedings and addresses, 1910, pp. 480-483.

75. ———. 1908. Journal of proceedings and addresses, 1908, pp. 385-447.

76. ———. Reports of the committee on six-year course of study. Journal of proceedings and addresses, 1907, pp. 705-710; 1908, pp. 625-628; 1909, pp. 498-503.

Treats of differentiation at the age of twelve.

77. New York State Education Department, Albany. Agriculture syllabus for secondary schools, 1907, pp. 37.

78. Owen, C. J Secondary Agricultural Education in Alabama. Office of Experiment Stations, bulletin no. 220, Nov. 19, 1909, pp. 30.

79. Ontario. Report of Inspection of the Agricultural Departments in the High Schools: being Appendix "AI" to the Report of the Minister of Education, 1907, pp. 952-966. [Reprint.]

80. Peet, C. E. What Shall the First-Year High-School Science Be? National Education Association. Journal of proceedings and addresses, 1909, pp. 809-816.

81. Poe, C. H. The Agricultural Revolution and the Teacher's Part in It. Rural Life Conference, 1909, Charlottesville, Va., pp. 72-83.

82. Rankin, Fred H. Developing the American Farm Boy: address delivered December 7, 1905. Published by the University of Illinois, pp. 26.

83. Robison, Clarence Hall. Administrative Phases of Agricultural Instruction; *in* proceedings of the conference on agricultural science, Amherst, Mass., July 1908. pp. 14-28.

84. ———. Some Text-books for Secondary School Agriculture. *Nature-Study Review,* vol. 3, no. 6, Sept. 1907, pp. 180-185.

85. Sanderson, E. Dwight. Biological Sciences in Their Relation to Agricultural Science. Proceedings of the second annual conference on agricultural science, Amherst, Mass., July 1909, pp. 50-58.

86. Seerley, H. H. National Aid in the Preparation of Teachers of Agriculture for the Public Schools. National Education Association. Journal of proceedings and addresses, 1909, pp. 965-968.

87. Snyder, Edward Reagan. The Legal Status of Rural High Schools in the United States. Columbia University Contributions to Education, no. 24, pp. 225.

88. Spalding, Edith H. The Problem of Rural Schools and Teachers in North America. Board of Education [London], Educational Pamphlet no. 13, pp. 70.

This represents an English point of view and is largely devoted to Canadian conditions.

89. STEVENS, F. L. The Farmers' Institute with Relation to Agricultural High Schools. The United States Department of Agriculture, Office of Experiment Stations, bulletin no. 213, pp. 53-57.
90. STEWART, JOSEPH S. First Annual Report of the Congressional District Agricultural Schools of Georgia. Bulletin of the University of Georgia, vol. 10, no. 4a, Dec. 1909, pp. 48.
91. STORM, A. V. Public School Agriculture. Iowa Yearbook of Agriculture, 1908, pp. 84-90.
92. THORNDIKE, EDWARD LEE. A Neglected Aspect of the American High School. *Educational Review,* vol. 33, March 1907, pp. 245-255.
93. ———. The Teaching Staff of Secondary Schools in the United States. Bureau of Education, bulletin no. 4, 1909, pp. 60.
94. TOWNSEND, E. J. The Status of the High School and Its Relation to Colleges and Universities. National Education Association. Journal of proceedings and addresses, 1909, pp. 794-799.
95. TRUE, A. C. Notes on the History of Agricultural Pedagogy in the United States. Proceedings of the 28th annual meeting of the Society for the promotion of Agricultural Science, 1907, pp. 84-106. [Reprint.]
96. ———. Progress in Secondary Education in Agriculture. Annual report of the Office of Experiment Stations, 1902, pp. 481-500. [Reprint.]
97. ———. Progress in Agricultural Education. Annual report of the Office of Experiment Stations, 1903, pp. 574-634.
98. ———. Progress in Agricultural Education. Annual report of the Office of Experiment Stations, 1904, pp. 576-616. [Reprint.]
99. ———. Progress in Agricultural Education. Annual report of the Office of Experiment Stations, 1905, pp. 303-357. [Reprint.]
100. ———. Secondary Education in Agriculture in the United States. Department of Agriculture, Office of Experiment Stations, circular no. 91, Nov. 8, 1909, pp. 11.
101. ———. What is Agriculture, Elementary, Secondary, and Collegiate? National Education Association. Journal of proceedings and addresses, 1908, pp. 1202-1207.
102. ——— and CROSBY, DICK J. The American System of Agricultural Education. Department of Agriculture, Office of Experiment Stations, circular no. 83, May 27, 1909, pp. 27.
103. UNITED STATES BUREAU OF EDUCATION. Benjamin Franklin's "Proposals Relative to the Education of the Youth in Pennsylvania," 1749. *Quoted in* circular no. 2, 1892, pp. 62.
104. ———. Report of the Commissioner of Education for the year ending June 30, 1908, pp. 740-741.
105. ———. Secondary Education *in the* Report of the Commissioner of Education for the year ending June 30, 1909, pp. 146-149.

105a. ———. Agricultural Education *in the* Report of the Commissioner of Education for the year ending June 30, 1910, pp. 255-278.

106. United States Department of Agriculture. *Experiment Station Record. In its* editorial, vol. 20, no. 10, May 1909, pp. 901-905.

107. ———. Federal Legislation, Regulations, and Rulings Affecting Agricultural Colleges and Experiment Stations. Office of Experiment Stations, circular no. 68, July 6, 1907, pp. 20.

108. ———. Institutions in the United States Giving Instruction in Agriculture. Office of Experiment Stations, Oct. 17, 1908, pp. 10.

109. ———. Institutions in the United States Giving Instruction in Agriculture. Circular no. 97, May 23, 1910.

110. ———. Organization, Work, and Publications of the Agricultural Education Service, corrected to December 1, 1909. Office of Experiment Stations, circular no. 93, Jan. 13, 1910, pp. 15.

111. ———. Report of the Committee on Extension Work, 1906-1907. Office of Experiment Stations, circular no. 75, Aug. 16, 1907, pp. 16.

112. ———. Secondary Courses in Agriculture. Seventh report of the committee on methods of teaching agriculture, of the Association of American Agricultural Colleges and Experiment Stations. Office of Experiment Stations, circular no. 49, pp. 10.

113. ———. A Secondary Course in Agronomy. Eleventh report of the committee on instruction in agriculture of the American Association of Agricultural Colleges and Experiment Stations. Office of Experiment Stations, circular no. 77, pp. 43.

114. Warren, G. H. Agriculture for High Schools. Proceedings of the second annual conference on agricultural science, Amherst, Mass., July 1909, pp. 32-43.

114a. ———. The Place of Agriculture in the Public High Schools. National Education Association. Journal of proceedings and addresses, 1910, pp. 1094-1103.

115. Waters, H. J. The Duty of the Agricultural College. *Science,* new series, vol. 30, no. 778, 1909, pp. 777-789.

Inaugural address at Kansas Agricultural College.

115a. Whitney, Worrallo. Relation of Pure and Applied Science to Secondary Education. *School Science and Mathematics,* vol. 10, no. 5, pp. 369-381, May 1910.

116. Wellington, Charles. Chemistry and Agriculture. *In* proceedings of the conference on agricultural science, Amherst, Mass., July 1908, pp. 35-43.

117. Willsey, Cora M. A Suggested Course in Agriculture. *In* papers and discussions by county normal training class teachers in Michigan. Published by the state superintendent of public instruction, bulletin no. 21, 1907, pp. 21-31.

## Topical List of References

[The numbers refer to the paragraphs of the foregoing list.]

# INDEX

(See also List of References, pages 191-200)